Library of Public Policy Administration

Volume 13

Series Editor
Michael Boylan, Department of Philosophy, Marymount University, Arlington, USA

Editorial Board
Simona Giordano, Reader in Bioethics, School of Law, Manchester University, UK
David Koepsell, Universidad Nacional Autónoma de México, Director of Research and Strategic Initiatives at Comisión Nacional de Bioética (CONBIOETICA) Mexico
Seumas Miller, Research Fellow, Charles Sturt University, Australia and Delft University, The Netherlands
Gabriel Palmer-Fernandez, Professor and Chair Philosophy, Youngstown State University, Youngstown, Ohio, USA
Wanda Teays, Professor, Philosophy Mount St. Mary's College, Los Angeles, CA, USA
Jonathan Wolff, Professor of Political Philosophy, Blavatnik School of Government, Oxford University, UK

Around the world there are challenges to the way we administer government. Some of these have to do with brute force that is backed by self-interest. However, there are those intrepid souls who think we are all better than this. This series of monographs and edited collections of original essays seeks to explore the very best way that governments can execute their sovereign duties within the sphere of ethically-based public policy that recognizes human rights and the autonomy of its citizens.

Proposals to the series can include policy questions that are nationally or internationally situated. For example, regional migration from victims of war, terrorism, police integrity, political corruption, the intersection between politics and public health, hunger, clean water and sanitation, global warming, treatment of the "other" nationally and internationally, and issues of distributive justice and human rights.

Proposals that discuss systemic changes in the structure of government solutions will also be considered. These include *corruption and anti-corruption, bribery, nepotism,* and *effective systems design.*

Series benchmark: 110,000-150,000 words. Special books can be somewhat longer.

More information about this series at http://www.springer.com/series/6234

Arthur J. Dyck

Achieving Justice in the U.S. Healthcare System

Mercy is Sustainable; the Insatiable Thirst for Profit is Not

 Springer

Arthur J. Dyck
The Divinity School
Harvard University
Cambridge, MA, USA

ISSN 1566-7669
Library of Public Policy and Public Administration
ISBN 978-3-030-21709-9   ISBN 978-3-030-21707-5 (eBook)
https://doi.org/10.1007/978-3-030-21707-5

© Springer Nature Switzerland AG 2019
This work is subject to copyright. All rights are reserved by the Publisher, whether the whole or part of the material is concerned, specifically the rights of translation, reprinting, reuse of illustrations, recitation, broadcasting, reproduction on microfilms or in any other physical way, and transmission or information storage and retrieval, electronic adaptation, computer software, or by similar or dissimilar methodology now known or hereafter developed.
The use of general descriptive names, registered names, trademarks, service marks, etc. in this publication does not imply, even in the absence of a specific statement, that such names are exempt from the relevant protective laws and regulations and therefore free for general use.
The publisher, the authors, and the editors are safe to assume that the advice and information in this book are believed to be true and accurate at the date of publication. Neither the publisher nor the authors or the editors give a warranty, express or implied, with respect to the material contained herein or for any errors or omissions that may have been made. The publisher remains neutral with regard to jurisdictional claims in published maps and institutional affiliations.

This Springer imprint is published by the registered company Springer Nature Switzerland AG.
The registered company address is: Gewerbestrasse 11, 6330 Cham, Switzerland

Introduction

Nothing can make injustice just but mercy. (Robert Frost: <u>A Masque of Mercy</u>, 1947[1])

Yes, good reader, the whole world is so affected and involved in this accursed avarice, fraud, false practice and unlawful means of support… that I do not know how it could get much worse. Yet they…call this … doing justice to all. (Menno Simons: <u>The Thirst for Profit</u>, ca' 1541[2])

In the concluding chapter of a book I published in 2005 on responsibilities and rights, I addressed the question of what it would take for the American healthcare system to achieve justice in providing medical care to its citizens and residents.[3] I never considered that the words of Robert Frost and Menno Simons quoted above would be remotely applicable to the US healthcare system. How surprisingly uninformed I turned out to be! I was surprised because there are so many admirable, highly skilled, and morally conscientious physicians, nurses, and other caregivers in the USA. In fact, I taught and met with many such at Harvard. I never imagined them to be laboring in a medical system that one notable, much published physician, Nortin Hadler, has declared to be "ethically bankrupt."[4] Hadler and other physicians, whose works I have cited extensively, have uncovered and documented the ethically and scientifically unjustifiable practices that plague the US healthcare system. Understandably, physicians are generally unaware of being involved in any such practices when indeed they are. One notable example of why this is so arises

---

[1] Robert Frost, "A Masque of Mercy," in *The Poetry of Robert Frost: The Collected Poems, Complete and Unabridged*, Edward Connery Latham, ed. (New York: Holt, Rinehart, and Winston, 1979), 521.
[2] Menno Simons, "The Thirst for Profit," in *Readings from Mennonite Writings: New & Old* (Intercourse, PA: Good Books, 1972), 372.
[3] Arthur J. Dyck, *Rethinking Rights and Responsibilities: The Moral Bonds of Community* (Washington, D.C.: Georgetown University Press, 2005), 280–326.
[4] Norton Hadler, *The Citizen Patient: Reforming Health Care for the Sake of the Patient, Not the System* (Chapel Hill, NC: The University of North Carolina Press, 2013), 5. Note: In this Introduction, I am not footnoting references to what appears in the various chapters of the book and is footnoted there.

from the scientifically unsound research, financed by the manufacturers of drugs and medical technology, that is being published in leading medical journals and approved by the Food and Drug Administration (FDA). Physicians expect to be correctly guided by this research: yet, in too many instances, they are not.

Before I engaged in the research for writing this book, I was unaware of these kinds of injustices within the US healthcare system. What then prompted me to write this book? First of all, in 2005, I argued that justice in the provision of medical care can only be attained if everyone's genuine needs for existing medical services are met regardless of their ability to pay sufficiently or at all for any services received. Justice as I understand it then and now is in accord with the traditional, now embattled, professional medical ethic. That moral guide for physicians has been in effect since at least the time of Hippocrates and his loyal followers. Hippocratic physicians denied no one in need of medical care, sharing their own resources to accommodate those who could not fully or at all pay for it. They also avoided harming their patients as much as possible. All of their interventions were based on careful and extensive observations, and because of this, they are praised for putting medicine on a more predictable scientific footing.

What I was aware of was that justice in accord with the professional medical ethic was presently, given the costs of medical care, increasingly often impossible for individual physicians to achieve. Furthermore, the Affordable Care Act, passed in 2010, does not fully cover all medical needs. At the same time, there was and is a widespread tendency among scholars to consider it just to assure universal access to medical care but only for receiving the so-called basic, minimum care. For some, it would be just to ration care. In both these instances, denying some medical care is being regarded as just, and the traditional medical ethic is being effectively ignored or rejected. As far as I am concerned, the very concept of what justice demands of all of us was and is under siege.

For all of the reasons cited, I decided that a book with a focus on justice and its demands in the way of providing people with medical care seemed not only worthwhile but necessary. Besides, some recent insights on the nature of our moral perceptions and motivations from the neurosciences would help make a stronger case for the traditional medical ethic than I was able to make in 2005.

In 2005, I made some modest suggestions as to how some of the costs of medical care could be reduced so as to meet everyone's necessary medical care. In the light of the current emphasis on limiting the amount of medical care in the name of justice, I knew I had to revisit the question of whether the traditional medical ethic is financially feasible. What I discovered was that with the ever-increasing costs of medical care, how the current system functions is not financially sustainable; despite its title, the Affordable Care Act is unaffordable. However, I also discovered that the mercy that changes injustices into justice is not only financially sustainable but will also improve the health and well-being of the patients who obtain care guided by mercy.

With the focus on justice, I begin and end Chap. 1 with an account of justice that I trace back to Hippocrates and his many followers. Their tenets and practices, as I noted above, are at the core of the traditional medical ethics guiding physicians to meet the demands of justice as I portray them in Chap. 1.

Chapter 1 draws upon the discoveries of the neurosciences that link mercy and justice to the empathic emotions. These are sources of knowledge of what mercy expects of us and sources of our motivation to act compassionately, that is to say, mercifully toward those who are suffering in some way. These motivational and cognitive capacities with which we are naturally endowed undergird and render actual the commitments and practices at the core of the traditional medical ethic, namely, denying no one medical care they need and avoiding harm and the risk of harm as much as possible: mercy and justice are at one in specifying what we owe one another as human beings.

Chapter 2 identifies certain processes and criteria by means of which we determine as nearly as possible that a given action or policy that we perceive and judge to be the one morally demanded of us can reasonably be defended as the one that is actually demanded of us.

With Chaps. 1 and 2 in hand, I had a cognitive basis for examining whether the claim that it is just to limit medical care for everyone to some basic minimum is justifiable. In Chap. 3, I found that those making this claim do not consider all of the moral demands of justice. The one they neglected is the moral demand to avoid harming one another, a key responsibility called for in traditional medical ethic. Acting on this responsibility would have the effect of limiting medical interventions, many of which are also costly. However, the traditional medical ethic also regards meeting the genuine medical needs of everyone as a demand of justice. Since I affirm this view of what justice demands, I was left with the task of finding convincing evidence that medical decisions made in accord with this ethical outlook are financially sustainable.

Well, by undertaking this task, I was in for some big surprises. First of all, I had no idea that an enormous amount of money is being spent and needlessly wasted on activities and medical interventions taking place in the US healthcare system that are not beneficial, and often harmful, and so cannot be defended ethically and scientifically. That troubles me rather deeply. However, the very existence of these very expensive practices means that there appears to be more than enough money that can be saved to stave off the present path to insolvency. Though it is not difficult to justify saving money in this way, getting the job done may prove to be difficult with all the money that is being gleaned from the illicit practices in question. And there are some additional ways to reduce the costs of medical care. Suggestions as to what ought to be done to achieve solvency for the US healthcare system are what Chap. 7 is all about.

Chapter 4 begins by addressing the unsustainable trajectory of the US healthcare system; historically, its expenditures grow at a rate that exceeds income by more than two percentage points. Furthermore, the percent of the gross national product (GNP) consumed by healthcare spending in the USA is much greater than any other nations. You might think that leading all other nations in what it spends for healthcare would make the USA the leader in health benefits derived from its healthcare system. Think again! Among a number of startling statistics cited in Chap. 4, consider this one: in 2010, the USA ranked 49th in the world for male and female life expectancy combined. And this is due to what is happening in the US healthcare system and not to other actions that affect life expectancy. The rest of Chap. 4 documents some of the medical practices the USA could and should change. The focus in this chapter is on overdiagnosis. Overdiagnosis occurs when symptom-free individuals are diagnosed as having conditions purported to require treatment though these conditions will never cause symptoms or death. These overdiagnosed individuals are receiving no benefits: they can only be harmed by the drugs usually being prescribed. Curbing these practices would save hundreds of billions of dollars annually and yield a considerable health benefit, particularly if symptom-free individuals receive scientifically sound advice regarding what constitutes a health-producing lifestyle.

Thankfully, traditional medical ethic has not died! Chap. 4 points to the growing recognition in the medical community that there are far too many costly medical interventions that cannot be clinically and scientifically justified. In addition to the physicians I cite in this book, Choosing Wisely is the name of an organized effort of a rapidly expanding number of medical societies dedicated to set standards aimed at avoiding medical interventions that can harm people without benefiting them and thereby avoiding unnecessary expenditures as well.

Chapter 5 is also concerned with overdiagnosis. In this instance, it takes place by treating normal ordinary behavior and feelings as mental disorders and thus medicalizing them. One way this happens is by expanding criteria for various categories of mental disorders featured in deceptive drug ads. As examples of this kind of diagnostic inflation, note these incredible increases in the diagnosis of the following mental disorders in the span of 15 years: childhood bipolar disorder, 40-fold; autism, 20-fold; a tripling of attention deficit/hyperactive disorder; and a doubling of adult bipolar disorder. These instances of overdiagnosing and medicalizing normal behavior can only occur when the criteria for these disorders are not followed. These conditions, the treatments of them, and the harms that result are discussed in detail in the said chapter. Clearly, diagnostic inflation has to be widespread and out of control when studies report very high prevalence rates of mental illness in the USA like this one: a study claiming that 80% of adults at age 21 meet the criteria for a mental disorder!

Chapter 6 examines quite a number of unethical practices carried on in the US healthcare system. When I refer to practices as unethical, I am identifying their failure in one or more respects to meet the demands of justice. Therefore, as an aid to the reader, I begin Chap. 6 by reviewing the demands of justice as depicted in Chap. 1.

Chapter 6 documents how drug companies trigger a great deal of the overdiagnoses and misguided drug treatments taking place in the US healthcare system. Among these unethical practices are the following: conducting, financing, and publishing rigged research; publishing studies that favor a drug's usage and suppressing those that do not; widely disseminating deceptive ads that falsely expand the symptoms a drug should allegedly treat; and developing financial ties to medical experts who develop guidelines for what conditions are to be treated with drugs and what drugs to use, thereby influencing them to expand the conditions that are to be treated with drugs manufactured by the companies with which they have financial ties. This has resulted in greatly increasing the number of people who are considered ill for a considerable number of diseases.

Another grossly unethical activity perpetrated by pharmaceutical manufacturers is that of charging unjustifiably high and ever-increasing prices for their drugs. The cost of drugs in the USA is a totally unnecessary, unjustifiable expense. These same FDA-approved drugs sell profitably for half as much on average in European industrialized countries. Inexcusably, the Affordable Care Act does not call for negotiating drug prices; drug company lobbyists prevailed!

Chapter 6 reveals the rather extensive ethical failures of the FDA whose mission is as follows: to assure the safety of medicines and medical devices, to assure that scientific information is sound and publically available, and to devise ways to keep down medical costs. Fully carrying out that mission would virtually, if not entirely, prevent the unethical conduct of medical manufacturers and much of the improperly misguided diagnoses and treatments in the US healthcare system.

Chapter 6 also examines what some hospitals do to keep the cost of healthcare in the USA at prohibitive levels indeed. One such example discussed details treating a patient for pneumonia at the University of Texas Southwestern Medical Center that led to a bill 161 pages long and 474,064 dollars high! The patient's insurance paid the hospital a lot of money – 71,169 dollars – but that left a bill of 402,955 dollars. The patient could only get the bill reduced to 313,000 dollars. The best offer of the hospital at the time this was reported was to cut the bill to 200,000 dollars if paid immediately off or the full 313,000 dollars if paid in monthly installments. The inflated charges for services are what make some nonprofit hospitals exceedingly profitable.

Charging more than even a good insurance policy will not cover harms to many people severely. Of the 65% of personal bankruptcies that involved illnesses or medical bills, 69% of those bankrupt for medically related reasons had health insurance when they filed. And grossly inflated charges for services prompt insurance companies to increase their premiums, deductibles, and co-pays.

Chapter 6 documents still another source of unjustifiably harming patients and driving up costs, namely, defensive medicine. Here is how one hospital CEO defends defensive medicine:

> We use the CT scan because it's a great defense... For example, if anyone has fallen or done anything around the head – hell, if they even say the word head – we do it to be safe. We can't be sued for doing too much.[5]

Three-fifths of American physicians acknowledge that they engage in more diagnostic testing than they regard as necessary because they fear being sued for malpractice. One study estimates the cost of defensive medicine as 45.59 billion dollars annually. Additionally, malpractice insurance is financially burdensome enough to

---

[5] Steven Brill, "Bitter Pill: Why Medical Bills are Killing Us," *Time*, March 4, 2013, 24.

cause some physicians to give up practicing medicine. The absence of tort reform in the Affordable Care Act is one more reason for its unsustainability.

Chapter 7 ends the book with suggestions about what can be done to put the US healthcare system on the path to sustainability, better medical care, and compliance with the demands of justice as spelled out and supported in the first three chapters of this book. The suggestions being made are almost exclusively based on the data and policies suggested by the physicians I cite. I am open to any policies that may more effectively curb and eliminate the injustices I have called attention to and documented in this book.

As the reader will discover, my book does not attempt to provide anything close to complete the coverage of the scope and types of notable activities that transpire in the US healthcare system. The focus is on some practices and policies that are unjust, that is, out of compliance with the traditional professional medical ethic. I remind the reader that this ethic expects all those involved in the care of patients, which includes the manufacturers of medicines and medical technology, to share their resources with patients, to sponsor, advocate and publish as medical knowledge only what is scientifically sound, and to pursue mercy for everyone in need of medical care and not overwhelm mercy by the relentless pursuit of profit and compensation.

Despite the existence of injustices in the US healthcare system, there is so much going on that is highly commendable. Lives that could not be saved in times gone by can now be saved by remarkably skilled physicians and by highly efficacious drugs. Medical devices can do so as well, and they can also aid individuals whose bodies have been injured in various ways to function normally or near normally to astonishing degrees. Many serious illnesses and epidemics of infectious diseases can be prevented by vaccines. The list of medical accomplishments is much longer than that and growing.

However, my book is sounding an alarm: to be sustainable and more health-promoting, the skills, medicines, and medical devices should be much more used in accord with sound medical knowledge, and the cost of these should be determined not only by profitability and compensation alone but also by mercy. Chapter 7 sketches some policies that could, if implemented, move the US healthcare system substantially toward achieving those goals. As Robert Frost so wisely observed, "Nothing can make injustice just but mercy."

Acknowledgments

I owe Lenny Lopez, M.D., a tremendous debt of gratitude. After all, while an assistant professor at Harvard Medical School, this former student of mine came to me with a proposal: publish a book that depicts justice. Lenny and his colleagues believed that there was a need for a work that made a convincing case for what a just healthcare system should look like. I soon discovered that such a case was needed.

I am exceedingly grateful for the support and assistance I received from William Reichel, M.D. He alerted me to books and articles in the medical literature and sent some to me as well. We became friends when this very thoughtful man volunteered his services at Harvard for several years to help me teach medical ethics. I learned a great deal from him.

I met my longtime friend, Rev. Joe Bassett, at Harvard—when he was a student and I a first-year assistant professor. He has been an enthusiastic and intellectually insightful supporter of my book. He was the one who suggested that I find a way to indicate spiritual roots. That led me to quote Menno Simons and Robert Frost. Thank you, Joe!

How can I possibly thank David Y. Kim enough for all his work on my behalf. This former student and now professor at Colgate Rochester Crozer Divinity School provided me with research and administrative support for completing the book. His reflections on the manuscript in its several versions were extremely helpful. I owe him so very much. He urged Springer to publish the book!

To Michael Boylan, the general editor of the series that is publishing this book, I owe a considerable debt of gratitude. He embraced the manuscript of my book enthusiastically and skillfully saw to it that it received a prompt, positive response from the publisher. He is not only very knowledgeable in the disciplines employed in the book, but he is also a very pleasant, helpful, and reliable person to work with. Thank you Michael!

It's wonderful to have friends and a family as beloved friends. My daughters, Cyndi and Sandy, put their computers and phones to work providing me with the information I needed about publishers and materials. The whole family was enthusiastic about the book—my dear wife, Sylvia, my daughters, and my grandchildren, Joseph, Jenny, and Jesse. They were willing to listen to what I was learning.

Notes
- Throughout Chaps. 4, 5, 6 and 7, I have footnote references to the books of a number of physicians. These references in turn provide the reader with extensive footnotes to the documentation of research in medical literature. I am thus inviting the reader to consult these books for what they have to say and in that way gain access to any other medical literature they wish to consult.
- Some readers may be concerned about the references in the book that tend to be somewhat dated: far from it! If anything, the changes documented in the book as needed to save the US healthcare system have grown more urgent to implement. Consult, for example, Edward P. Hoffer' book, *Prescription for Bankruptcy: A Doctor's Perspective on America's Failing Health Care System and How We Can Fix It* (West Wareham, MA: Omni Publishing Co., 2018).

Contents

1	What Justice Demands...................................	1
2	The Cognitive Bases for Deciding When Policies Are Just.........	23
3	Advocating Basic Minimum Medical Care: A Case of Justice Denied.....................................	41
4	Overdiagnosing, Overtesting, and Overmedicalizing Physical Conditions......................................	83
5	Overdiagnosing, Overtreating, and Overmedicalizing Behavior and Feelings..	113
6	Practices and Policies in the U.S. Health Care System that Are Scientifically and Ethically Unjustifiable: They Should Not and Cannot Persist.....................................	135
7	Suggesting Policies and Practices for Increasing Justice and Assuring the Sustainability of the U.S. Healthcare System.....	167
	Bibliography..	205
	Index...	209

Chapter 1
What Justice Demands

Abstract Drawing upon the neurosciences and other empirical observations, I argue that the moral demands of justice are expressed in our natural inhibitions and proclivities, namely inhibitions of avoiding harming one another and in our proclivities to spawn offspring, nurturing and sustaining them, and protecting and rescuing their lives. These key moral demands undergird the two main conclusions of chapter one: healthcare ought to be universal; and that this goal is possible if the money to provide it comes from the profits generated by all individuals and entities that provide healthcare, and by avoiding unethically generated harmful medical practices.

"Have a good look at [the patient's] affluence and means, sometimes treating gratis…" That admonition addressed to physicians is in the "Precepts," part of the corpus of literature expressing the ideas and ideals of the Hippocratic tradition.[1] Physicians were praised in Hellenistic inscriptions for treating the rich and poor.[2] This practice and this ideal to treat even those who can pay little or nothing for their care is one of the hallmarks of the traditional medical ethic guiding medical practice since ancient times. Attesting to the historical continuity of this ethic is the ninth century medieval text, declaring that: "If the patient be wealthy, let this be the proper occasion for profit; if a pauper, any reward is sufficient for you."[3] We find this same ethical ideal asserted in the AMA Code of 1903 and persisting in the AMA Code of 1957, which states that: "His [physician's] fee should be commensurate with the service rendered and the patient's ability to pay."[4]

According to what we are calling the traditional medical ethic, no one should be denied medical care on the basis of cost or inability to pay for it. This moral impera-

---

[1] Oswei Temkin, Hippocrates in a World of Pagans and Christians (Baltimore: John Hopkins University Press, 1991), 30.
[2] Ibid., 219.
[3] Ibid., 223.
[4] Stanley J. Reiser, William J. Curran, and Arthur J. Dyck., eds., Ethics in Medicine (Cambridge: MIT Press, 1977), 39.

tive is not simply one to provide for universal *access* to care but for universal *care* as well. Furthermore, the care to be provided for those who can pay very little or nothing at all is that which everyone receives for the medical condition being treated. As the AMA Code of 1957 states, "His [physician's] fee should be commensurate with the service rendered and the patient's ability to pay." What justice demands, therefore, is an equal measure of medical care for all. What so many in the contemporary literature on health policy refer to as a "decent minimum," or as "basic benefits" for everyone, does not accord with the view of the traditional medical ethic, falling short as it does of what justice demands.[5] The same can be said of the AMA Code of 1980 which omits any reference to the obligation to provide care even to those who can pay little or nothing for it.

The reader will rightly be thinking about how the circumstances in which medicine is conducted have greatly changed, and how the costs of providing medical services have skyrocketed. These changes are undeniable. The question before us, however, is how in the name of justice we respond to these changes. How we answer this question depends, first of all, upon our ability to identify the demands of justice, particularly those that pertain to the provision of medical care. The argument put forth in this chapter is that a proper understanding of justice undergirds and justifies nothing less than fully accorded care for all, regardless of ability to pay fully or at all, as called for by the traditional medical ethic. Those who favor universal access to medical care but limit care to a "decent minimum," are denying care to those who cannot afford the care they need that is not included in the provisions for a "decent minimum" of care. That policy overlooks the very moral imperatives that move them, and all of us, to be concerned about anyone whose illnesses go untreated for whatever reason. These moral impulses to meet the medical needs of everyone shaped the traditional medical ethic. Any theory of justice pertaining to the distribution of medical care should account for such a longstanding commitment to universal medical care. Chapter 3 will indicate why some prominent contemporary theories of justice fail to offer such an account, and fail as well to justify deviating from an unqualified commitment to universal medical care.

Our concern in this chapter is to identify what it is that justice demands of us. This account of justice provides a basis for the traditional medical ethics as we find it in the treatises and practices of Hippocratic physicians, a portrayal of which will follow what we have to say about justice.

---

[5] Chapter 2 will present, analyze, and critique the works of three very influential scholars who advocate a universally accessible decent minimum of medical care. They are: Norman Daniels, Just Health: Meeting Health Needs Fairly (New York: Cambridge University Press, 2008); Martha C. Nussbaum, Frontiers of Justice: Disability, Nationality, Species Membership (Cambridge, MA: Harvard University Press, 2006); and Amartya Sen, The Idea of Justice (Cambridge, MA: Harvard University Press, 2009).

1.1 Justice: What We Owe One Another

Justice is what we owe one another. In this, its generally accepted generic sense, justice refers to all of our most basic moral obligations: Those we owe others and others owe us. When we say that justice is what we owe others and others owe us, we are saying that justice is comprised of moral behavior we can expect and claim from one another as moral rights.[6] John Stuart Mill, the nineteenth century British philosopher whose thinking continues to emerge in deliberations over matters of public policy in the U.S. and many other countries, makes this same point: "Justice implies something that is not only right to do, and wrong not to do, but which some person can claim from us as his moral right."[7] Furthermore, to speak of rights is to speak of society's obligation to sustain the circumstances that actualize the moral demands of justice. As Mill would have it: "When we call anything a person's right, we mean that he has a valid claim on society to protect him in the possession of it, either by force of law, or by that of education or opinion."[8] A society that fulfills its obligations to affirm and guarantee human rights is a just one.

With this conception of justice in mind, there are at least two questions that should be addressed. First, what are these moral rights? Secondly, on what grounds can we validly claim that societies are obligated to protect human rights? Mill does address these two questions. Though his answers to these questions will need to be refined and further developed, the manner in which he seeks these answers suggest a way forward.

For Mill, justice has its origin in our natural proclivity "to repel or retaliate a hurt or damage to oneself" or one's fellow human beings.[9] This sentiment expresses itself in moral rules "which forbid mankind to hurt one another."[10] Among the harms proscribed by these rules, Mill includes: Wrongful aggression; wrongful exercise of power over someone; and wrongfully withholding something from someone.[11] These rules are binding in the sense that they specify behavior we have a right to expect and exact from one another. Injustices, such as wrongful aggression against us, are violations of our rights. Society, moreover, has the obligation to protect us from any violations of our rights.

Society's obligation to secure and protect our rights is something we can validly insist upon. That is the case, according to Mill, because the moral rules that comprise justice and specify our rights involve the most vital of all interests, namely

---

[6] Daniels, Just Health, I and II refers to justice as what we owe one another but as the discussion of this work in Chap. 2 will document, he applies a much narrower version of justice in his assessment of health policy.

[7] John Stuart Mill, "Utilitarianism," in The Philosophy of John Stuart Mill, Marshall Cohen, ed. (New York: Random House, 1961), 380.

[8] Ibid, 384.

[9] Ibid.

[10] Ibid., 391.

[11] Ibid., 392. Note that Mill includes good for good as one of the dictates of justice and argues that it is a grievous harm when good is denied. This he does on pages 392 and 293.

security: It is most vital because security is what "no human can possibly do without."[12] Furthermore, "this most indispensable of all necessaries, after physical nutriment, cannot be had unless the machinery for providing it is kept unintermittedly in active play."[13] Doing so, Mill asserts, "makes safe for us the very groundwork of our existence."[14]

Mill does make a strong case for some of the very important links between justice and medical practices. There are a number of ways in which a patient's right to be free from any injuries that are not sufficiently beneficial to be warranted is protected, ways such as: The injunction to "First of all do no harm"; the legal enforcement of the right to refuse treatment and to be treated only upon one's consent or that of a legal guardian; and legal redress for injuries resulting from malpractice. What Mill's portrayal of the sentiment of justice lacks is an unambiguously recognized proclivity to care for one another as a matter of justice, as something we have a right to. One can certainly argue that failure to care for someone who is ill is harmful to that individual. That, however, does not invest justice with any natural urge to offer such care and provides no clear basis for depicting medical care as a moral right to be supported societally. We need to go beyond viewing justice as only, or primarily, society's obligation to protect us from wrongful injuries or harms.[15]

Interestingly, Mill's argument for obligating society to protect moral rights includes this contention: Justice "makes safe for us the very groundwork of our existence."[16] This suggests how one might expand what justice demands of individuals and societies to include caring for others and thus putting a floor under the societal obligation to provide for medical care as everyone's right. Think of the "groundwork of our existence" as a whole set of human relations that are morally requisite to bring into being and sustain individuals and communities. Proceeding in this way, we can show that justice, what we owe one another, includes much more than the avoidance of harm, important as that is as one of the moral requisites for protecting individuals and maintaining communal life.

1.2 Justice: The Moral Requisites of Individual and Communal Life

Without the interpersonal relations that make communities possible, no individuals would exist; without the existence of individuals, there would be no communities. What are these requisites of communal life and how are they realized?

---

[12] Ibid., 385.

[13] Ibid.

[14] Ibid.

[15] Mill distinguishes justice and beneficence. We have no right to beneficence in his view (Ibid., 380–381). This distinction persists. See, for example, the widely used introduction to ethics by William K. Frankena, Ethics (Englewood Cliffs, N.J.: Prentice-Hall Inc., 1963).

[16] Mill, "Utilitarianism," 385.

Human beings exhibit and act upon certain natural proclivities and inhibitions. Were human beings devoid of any natural proclivity to have children, human beings, however they come to be, would cease to be and so, of course, would their communities. For individuals and communities to come to be and persist, it is not enough to have the proclivity to procreate. It is necessary as well to have the proclivity to nurture life—one's own, that of one's children, and that of others on whom one depends. Individuals cannot come into being and be sustained unless they are nurtured by those who have the wherewithal to do it, and the proclivity to take this on as a responsibility they feel obligated to carry out. From the earliest states of human development, no one survives without nurture. In a word, nurture is necessary to sustain and protect human lives.

To nurture a life is to take responsibility to assure that the necessities of life, such as shelter, nutrition, and a safe environment, are available and can be utilized by the one being nurtured. Nurture aims as well to prevent illnesses and care for those who are ill. Parents who fail to meet these obligations to nurture can be legally charged with child neglect and, in extreme cases of child neglect or physical abuse, have their children removed from their custody to be placed in a properly nurturing environment. Nurture is legally enforced as a right in such instances.

Society supports nurture in other ways and does so for everyone. Parents and individuals generally cannot meet their responsibility to themselves and one another, nor maintain well-being, without the assistance of others and of society. The obligation to nurture is expressed communally in a number of ways. Generally, societies that we know about, have provided care for those who are ill or injured, and have among their members, individuals recognized as healers and caregivers. Societies also furnish weaponry and create armed forces to protect life against aggression and assaults upon life and property. Contemporary communities also protect life and property by way of fire and police departments. Communities and their individual members expect that all who serve in these various professions will not only protect lives but also rescue the lives of those in peril, whether by reason of injuries, illnesses, accidents, fires, assaults, or natural disasters. In keeping with their expectations, these professionals see to it that the means are provided, and the duty to rescue can be successfully carried out.

That nurture includes the duty to rescue is clearly evident in parental behavior. Anyone who reads a daily newspaper will soon enough learn of the more dramatic instances in which parents, at great risk to themselves, have rescued or attempted rescue one or more of their children. Sometimes these exploits result in serious injuries, sometimes in death, for the parent or parents involved.

However, heroic effort at rescuing people is not exclusively parental behavior. Newspapers also report the risks taken by individuals, acting alone or in concert with others, to save the lives of strangers, whether adults or children. These reports include the risks of entering burning buildings, pulling someone from a car engulfed in flames, and jumping into an icy, swift stream to save someone from drowning. As I write this chapter, these are among some recently reported events. It is noteworthy that these people when interviewed, usually commented that they had only done, either what anyone else would do, or what anyone ought to do, in such circumstances.

They gave the duty to rescue the strongest endorsements, coupled with a considerable exhibit of humility: They did not see themselves as going beyond the call of duty.

At the same time, our concern for human life extends also to any would-be rescuer. We do not expect anyone to take undue risks to life and limb, or to take actions reasonably deemed to be futile. This suspension of the duty to rescue in situations reasonably judged to be too risky or futile holds as well for firefighters, the police, and physicians. At the same time, because of their various skills and professional obligations our society has set up an arrangement such that calling 911 provides a way to bring the relevant professional skills to bear in circumstances in which someone's life depends upon doing just that. A failure to summon professional life-saving assistance when deemed necessary is regarded as a serious moral lapse. Our whole nation was shocked when the media reported that witnesses viewing a crime in progress from the safety of their apartments failed to call the police, and after some efforts to escape, Kitty Genevese was fatally stabbed to death. Given the time it took, the crime appeared to be one that might have been stopped.

There are in continental European legal systems criminal sanctions (fines) that are exacted from those who fail to rescue. The responsibility is a moral minimum, requiring only actions that do not involve significant costs or risks, actions such as summoning the appropriate authorities. Since these laws serve a mostly hortatory purpose, the fines are not excessive. As the legal scholar Mary Ann Glendon has observed regarding these laws,

> Continental European commentators take it for granted that making the failure to rescue a public wrong will operate to encourage compliance, with certain basic duties attaching to good citizenship. As a leading French legal scholar [André Tune] put it, the rescue laws serve as reminders that we are members of society and ought to act responsibly.[17]

In effect then, European law recognizes the responsibility to rescue as a requisite of community. By enforcing this responsibility, a moral and legal right to rescue is being acknowledged: It is a matter of justice to do so.

Comparable laws do exist in the U. S. Largely due to the urging of the medical profession, all American jurisdictions have enacted "Good Samaritan" laws that grant immunity from civil liability to individuals who unintentionally cause harm when they voluntarily aid an endangered person. When adopting such a statute in 1967, the State of Vermont added to its version a penal section that established a general duty to assist someone in peril. Later Minnesota passed a similar law. Having cited these laws, Glendon observed that, "there is nothing alien to American values in making the failure to come to the aid of an endangered person a criminal offense."[18]

We have portrayed some of the most significant ways in which nurture is a moral requisite of individual and communal life. Our society explicitly recognizes nurture as a right, a matter of justice, by instituting organizations, professions, and in some

---

[17] Mary Ann Glendon, Rights Talk: The Impoverishment of Political Discourse (New York: The Free Press, 1991), 84–85.

[18] Ibid, 88. Glendon is taking issue with the impression to the contrary created by certain court decisions and legal textbooks.

cases, laws to assure that human life is cared for, protected, and rescued in life-threatening situations. What fuels all of this is the natural proclivity to nurture human life, our own and that of others. We know this since we do in fact have functioning communities and would not if such proclivities in us were insufficient or utterly lacking for bringing about these requisite relations to one another.

By themselves, however, these various embodiments of the proclivity to nurture do not suffice to sustain and protect our individual lives and communities. To accomplish this, it is necessary as well that we as human beings have certain internal inhibitions that serve to restrain us from actions destructive of individual and communal life and well-being. Paramount among the harms we have inhibitions about inflicting upon one another are intentional acts of killing, stealing, and lying. Limiting these moral wrongs is so essential to maintaining every aspect of individual and communal life that we have laws precisely designed to do just that. Laws serve to reinforce and strengthen our inhibitions against killing, stealing, and lying. This they do by punishing violations of our moral rights to be free of these harms, and by publicly identifying and specifying concretely these rights.

It should be evident by now that just laws are also requisites of individual and communal life, functioning as they do to spell out specifically, and punish and deter the moral wrongs that are particularly destructive of individual and communal life. Whereas we generally rely on education and social disapproval to aid us in being inhibited about lying, there are instances in which we consider it necessary to do more. Thus, for example, to prevent deceit in fiduciary relationships, business transactions, contractual arrangements and proceedings in our courts, we punish those who deceive in such interactions, and where appropriate, compensate victims of such deceptions for any losses they have suffered. By similar means, laws seek to keep people from stealing and killing.

Laws also provide moral instruction. Intentionally killing an innocent person is a punishable wrong. At the same time, the law recognizes the preciousness of human life by condoning killing in self-defense when the circumstances warrant it, that is, when there is apparently no other immediately discernible way to save one's life from an aggressor. In this instance, the duty to rescue oneself in this way is being legally invoked and condoned. The preciousness of human life is acknowledged also by the severity of the penalties for murdering someone.

One should note that the protection of human life is not confined to the laws against murder. There are regulations and laws that protect life and health from various hazards such as those introduced into our air, water, food supplies, buildings, and workplaces.

Laws, regulations, and enforcement agencies, then, do for our communities and ourselves, what, for the most part, we cannot do for ourselves as individuals. They help keep us from being harmed, even to the point of incarcerating for life those who are judged to be a continuing threat to our very lives were they to be released. Laws are in certain respects necessary but not in themselves sufficient for sustaining individual and communal life and well-being. Laws would neither exist nor achieve their purposes were we as human beings bereft of our proclivities to act justly, and our inhibitions about acting unjustly. Is there evidence that these natural proclivities and inhibitions exist? Yes! Recent work in the neurosciences and contemporary psychology yield evidence confirming the existence and nature of our natural proclivities and inhibitions as guides and motivating forces to be moral, to be just, in the ways we have been describing.

1.3 Empathy: The Source of Our Natural Propensities for Justice

As human beings, we have the natural capacity to be moral. "Empathy" is the key term neuroscientists use to refer to that capacity.[19] Based on the discoveries of the neuroscientists, we know that empathy, since its functions depend upon being possessed of a human brain, is a shared propensity. It is empathy that enables us to know what is just and be motivated to behave justly.[20] Based on neuroscientific research as well, we know that the loss of certain areas of the brain render individuals incapable of making moral decisions of the kind that meet the moral obligations that ordinary life demands and expects of us, such as being reliable enough to hold a job, or to predictably obey the laws of the land.[21] What such individuals lack are the emphatic emotions that would lead them to care about themselves and others, emotions essential for being morally responsible.

In the context of conducting their research on empathy, neuroscientists and psychologists have defined empathy in the following differing ways: "(a) *knowing* what another person feels; (b) *feeling* what another person is feeling; and (c) *responding* compassionately to another person's distress."[22] The ability and will to engage in acts that accord with the demands of justice depend upon being empathic in all three ways described in these definitions. To respond in a morally appropriate way to someone suffering or otherwise in need of assistance, one has to be able to discern accurately the distress of that individual (empathic accuracy), one has to feel some of the distress being felt by that individual (sympathy), and one has to feel the urge to help (compassion). When we speak of acting justly as informed and fueled by the empathic emotions, the expression "empathic emotions" refers to all three expressions of empathy, namely "empathic accuracy," "sympathy", and "compassion."

---

[19] See, for example, Allan N. Schore, Affect Regulation and the Origin of the Self: The Neurobiology of Emotional Development (Hillsdale, N.J.: Lawrence Erlbaum Associates, 1994), 348–354.

[20] See, for example, Martin L. Hoffman, Empathy and Moral Development: Implications for Caring and Justice (New York: Cambridge University Press, 2000). Note that Hoffman distinguishes caring and justice but does portray empathy as the source of our knowledge of both and of our motivation to care and to behave justly.

[21] Atonio Damasio, Decartes' Error: Emotion, Reason, and the Human Brain (New York: Penguin Books, 2005), 379. See also, Schore, Affect Regulation and the Origin of the Self, 351–354.

[22] Robert W. Levinson and Ann M. Ruef, "Empathy: A Physiological Substrate," Journal of Personality and Social Psychology 6:2 (1992), 234.

1.3 Empathy: The Source of Our Natural Propensities for Justice

In his very extensive review and interpretation of scientific research pertaining to what he calls the "neurobiology of emotional development," Allen Schore includes a chapter on early moral development.[23] He indicates that "an emphatic reaction to the distress of another" arises early in life at 10–14 months, and that the capacity for empathy is "an essential prerequisite to alter social and moral development..."[24] Already at 18 months, children exhibit "moral prosocial altruistic behavior in the form of approaching persons in distress and initiating positive, other oriented, affective and instrumental activities in order to comfort the other."[25] As an immediate result of this empathic, altruistic response to the distress of another, children who comfort someone else experience a decrease of their own distress.

This means that at the age of 18 months children are able: To perceive how someone else feels; feel pained by someone else's pain; and feel an urge to help someone whose pain they have perceived and felt. In short, they have the capacity for empathic accuracy, sympathy, and compassion. These empathic emotions inform, guide, and motivate our propensities for acting justly. They also inhibit us from inflicting injuries or harming others. Citing some literature to that effect, Schore notes that: "other-oriented empathy...mediates the important function of inhibiting interpersonal aggression..."[26] Thus the proclivity to act justly and the inhibitions about acting unjustly are naturally and universally present in us as human beings, rooted as they are in our brain's capacity for being empathic.

Moral development has social as well as biological roots. Schore cites a number of studies indicating that there is an interaction between the development of the brain and the nature of the care being provided to infants.[27] Though caregivers influence the development of the brain, the character and extent of that influence depends upon what the brain is capable of doing at its various stages of development. Furthermore, if infants, children, or adults experience sufficient damage to the right hemisphere of their brain, particularly of the orbital frontal lobe of the cerebral cortex, they will lack the empathy to discern and do what is morally right.[28] Schore summarizes the research on the social and biological underpinnings of our moral nature as follows:

> The caregiver influences the trajectory of the child's developing moral capacities by shaping the neurological structural system that mediates such functioning. Neuropsychological and neurological studies suggest that the orbitofrontal cortex is centrally involved in empathic and moral behavior. Structural deficits in this system are associated with sociopathic pathologies.[29]

---

[23] Schore, <u>Affect Regulation and the Origin of the Self</u>, 348–354.
[24] Ibid., 350–351.
[25] Ibid., 351.
[26] Ibid.
[27] Ibid., 352–353.
[28] Ibid.
[29] Ibid., 354.

Caregiving, of course, takes place within diverse cultural contexts. This well-studied phenomenon has led many scientists and philosophers to regard morality as purely socially constructed and culturally relative.[30] Such a view overlooks the fact that cultures can only exist if individuals and communities exist, and these can only exist if the natural proclivities and inhibitions exist to actualize what is morally requisite for there to be individual and communal life. What is more, thinking of moral rules and behavior as totally culturally relative is no longer tenable in the light of what we are learning from the neurosciences. That is the conclusion the psychologist Martin Hoffman comes to in his explicit analysis of cultural relativism.[31] As he observes, the capacity for empathy resides in our brains and hence empathy, and the compassion and justice it makes possible, occur as universal, cross-cultural phenomena.[32]

Laws reflect the universality of justice as we have depicted it, existing as they do in some instances, to protect the health, life, property, contractual arrangements, and commercial transactions of individuals and groups within their jurisdiction. They exist as well to care for and assist individuals and groups whose needs cannot be met without assistance. In these ways, they lend support to the moral requisites of individual and communal life, and to the proclivities and inhibitions from which they, and also laws, originate and continue to be actualized.

In their specificity, laws vary. Unlike the natural sources of the moral demands of justice, laws are constructed, designed to meet the goals of justice by specific means, suitable for specific circumstances. Laws governing vehicular traffic in Massachusetts, for example, differ from those obtaining in New Hampshire. Yet, justice, in this case safe travel, is served to the extent that these laws are reasonable, made known, and obeyed. Making laws is not an exact science. Laws may need to be adjusted because of inadequacies that show up, or circumstances that change.

Like laws, all our particular moral decisions and policies, whether private or public, are relative to the situations within which they occur and to which they apply. However, the moral requisites of individual and communal life, the moral demands of justice, are universal, referring as they do to the right and wrong-making characteristics of any laws, decisions, or policies described as just or unjust. Our specific laws, decisions, and policies are usually morally complex in the sense that they involve more than one right or wrong-making characteristics.[33] Thus, for example, killing is always wrong-making but self-defense may require an action intended only to stop a killing which may necessarily or inadvertently be an act of killing someone. Yet we can regard this act as just since it saves a life unjustly attacked. Similarly, surgery has the wrong-making characteristic of subjecting an individual

---

[30] For a detailed analysis of relativism, see Arthur J. Dyck, On Human Care: An Introduction to Ethics (Nashville, Tenn.: Abingdon, 1977), 114–134.

[31] Hoffman, Empathy and Moral Development, 273–283.

[32] Ibid., 274. See also Marc D. Hauser, Moral Minds: How Nature Designed Our Universal Sense of Right and Wrong, (New York: HarperCollins Publishers, 2006).

[33] These observations were inspired by the distinction between *prima facie* duties and duty proper found in W. D. Ross, The Right and the Good (Oxford: Oxford University Press, 1930), 16–47.

to the risks of suffering and death. However, it can be just: If the probability of death or continued suffering is much greater than being without surgery; if there is no known safer alternative method to prevent death or relieve such suffering; if the individual consents to this particular surgical intervention; and if the surgeon is appropriately qualified. For children and some others, consent may require a guardian. We can think of the moral requisites of individual and communal life as the moral elements that enter into our deliberations as we seek to realize justice through our particular laws, policies, moral guidelines and behavior.

The discoveries being made in the neurosciences provide a significant empirical basis for asserting the universality of our natural proclivities to act justly and our natural inhibitions restraining us from acting unjustly. Since we share these proclivities and inhibitions as human beings, we expect others to seek justice and shun injustices as we are inclined to do. Moral rights, then, exist in the form of expectations that we will, or ought to be, treated justly by others and not be subjected by others to injustices perpetuated against us. That means that everyone can justifiably claim to have a moral right to be treated justly. We cannot predictably secure all of our moral rights individually. Therefore, our claims to justice include some moral claims upon our communities to help us realize certain of our moral claims. For that purpose, members of communities engage in cooperative and collective actions that create and sustain the appropriate moral rules, institutions, regulations, and laws as deemed necessary. By these means, just relations among us are more extensively and more predictably attained, and the moral requisites of our individual and communal life maintained.

We have identified some of the fundamental moral imperatives of justice. We say "some" because we do not wish to claim that such a brief characterization of justice has yielded an exhaustive list of moral responsibilities and rights that may properly qualify as universally existing moral requisites of individual and communal life. However, by having recourse to a concept of justice in its generic sense, that is, as what we owe one another as human beings, we have uncovered the moral imperatives of justice that can be discerned in the literature and practices of Hippocrates and Hippocratic physicians, and that have served to guide physicians for centuries. We turn now to consider the key moral ideals found in the Hippocratic tradition, ideals that have sparked so much emulation by physicians ever since their incarnation by Hippocrates.

1.4 Justice in the Hippocratic Tradition

In his masterful portrayal of the Hippocratic tradition in the ancient world, the physician and historian of medicine, Owsei Tempkin, calls attention to a moral dilemma physicians faced then and still face now in still greater measure. Tempkin describes how this dilemma arises and is confronted in the instruction that is offered in Precepts, one of the Hippocratic treatises.

Precepts urged that everything possible be done for the healing of the sick. But then, as now, the demand for complete dedication to the art conflicted with the doctor's need to make a living by its practice. How to resolve this conflict is a problem of medical ethics for which Precepts found its answer in adjusting self-interest to philanthropy.[34]

The author of Precepts does not believe that seeking and accepting a fee in any way demeans physicians but he advises them,

> Not to give way to excessive inhumanity…but to have a good look at [the patient's] affluence and means, sometimes also treating gratis…And when the occasion arises to support somebody who is an alien and without resources, such people in particular must be helped.[35]

These admonitions did not go unheeded! There is ample evidence that physicians in ancient times were honored for precisely carrying out what these moral imperatives expected of them. Many inscriptions have been found citing the virtues of physicians acting in accord with these Hippocratic moral injunctions. Consider this sample: An inscription praises a physician for serving all alike fairly, whether poor or rich, slaves or free or foreigner, and for maintaining "a blameless reputation in all respects, providing proper attendance, which was open to all, as befits a man of culture and moral sense."[36] Here you have an instance of practicing the ethic of universal medical care, turning no one away from needed care whatever their means, their status in society, or their nationality.

At the very heart of the Hippocratic ethic is the insistence that the proper practice of medicine requires physicians to be humane and that being humane will enhance the quality and efficiency of care. We now know that humaneness is motivated and informed by our empathic emotions. Indeed, the author of Precepts specifically lauds these emotions when he asserts that doctors who love people will be devoted and successful practitioners. In fact, the author of Precepts observes that such decency can in itself be therapeutic: "For some patients though aware that their condition is not without danger, go on to recover their health simply through their satisfaction with the decency of the physician."[37]

The author of Precepts extends the obligations of humane physicians by advocating that: "It is indeed well to take charge of the sick for the sake of their health and to give thought to the healthy for the sake of their freedom from disease, also [it is well] to give thought to the healthy for the sake of decorum."[38] At this point, we see the Hippocratic ethic emphatically committed, not only to every effort to restore the health of those who are ill, but also to assist healthy people to avoid illnesses and stay healthy, and to offer the welcome assurances experienced by those who are examined and declared to be healthy.

---

[34] Oswei Tempkins, Hippocrates in a World of Pagans and Christians, 29.
[35] Ibid., 30.
[36] Ibid., 20.
[37] Ibid., 31.
[38] Ibid.

1.4 Justice in the Hippocratic Tradition

The Hippocratic ethic commends empathy in yet another critically important regard: "To help or not to harm."[39] First of all do no harm has been understandably inferred from the Hippocratic treatises and the behavior of physicians guided by them. Applying this maxim is critical for practicing medicine justly, for doing so shields patients from medical interventions that are only harmful or more harmful than beneficial; and that saves money that would otherwise be ill spent. At the same time, the quest to avoid unjustifiable harm to their patients motivates physicians to make more use of alternative therapies that are scientifically sound and essentially benign. These include therapies similar in kind to some promoted and used by Hippocratic physicians, such as good and appropriate nutrition and healthy habits.

It is important to realize that the Hippocratic treatises insisted that diagnoses and treatments be based on data, on careful observations and factual information, and not on speculative theorizing. W. H. S. Jones has thoroughly studied and translated into English the Hippocratic corpus and he regards the Hippocratic tradition as the starting point of a scientifically oriented outlook on how medicine is to be practiced and improved. He summarizes this aspect of the tradition when he characterizes the Hippocratic treatises as "embodying a consistent doctrine of medical theory and practice, free from both superstition and philosophy, and setting forth rational empiricism of a strictly scientific character."[40] Being scientific in deciding when and how harm to patients can and should be avoided must be linked if justice is to be attained. Why this is being stressed and underlined stems from the unfortunate fact that both the maxim of first doing no harm, and of practicing medicine in accord with trustworthy scientific data, are not as uniformly and widely occurring as one might well be led to assume. These erosions of justice are extremely costly. They take the form of excessively and inappropriately resorting to the use of expensive drugs where, for example, the adoption of a healthy lifestyle would be much more effective and free of harmful side effects: And there is ample sound science that supports precisely such moves. These matters will be quite thoroughly aired and documented later in Chap. 4. In a word, were the medical profession to heed much more extensively the demands of justice Hippocratics taught and acted on, patients would greatly benefit and medical costs would markedly decrease.

These Hippocratic ideals comport with justice. The moral imperatives of justice include the natural urges to care for and protect one another's health and life and when necessary, to rescue one another from life-threatening situations. These are expressions of justice as nurture. Human beings normally share the empathic emotions that fuel this natural proclivity to nurture. Hence we reasonably expect that physicians will have these same empathic urges to help, care for, and rescue us as the need arises. Given these empathic drives, it is no surprise that the Hippocratic ethic arose as a guide for physicians. That ethic regards all patients as having a moral right to the medical care available and appropriate to meet their needs.

Empathy also inhibits us from injuring one another. Patients expect, and physicians pledge, that they will first of all do no harm. The universal nature of this

---

[39] Ibid., 28.
[40] Jones, W. H. S., Hippocrates (Cambridge: Harvard University Press, 1923), Vol. I, xxi.

inhibition makes trust possible, the trust that any risk of injuries from medical intervention must and will be justified by a therapeutic rationale and a reasonable chance of success. This same inhibition about causing harm, makes it as gratifying for physicians as it is for patients whenever physical examinations are free of negative results.

There are, then, strong natural impulses that guide medical practices and our expectations with respect to what they entail. Therefore, from a moral perspective, medical care for everyone requiring it has always been a moral ideal. For centuries, physicians have generally and voluntarily sought to make this a reality.

However, the context in which medical care is offered and received has changed markedly. Whereas the AMA Code of 1957 still called upon physicians to provide fully for everyone's care, that injunction was dropped from the AMA Code of 1980. The sheer cost of medical services alone undermined any realistic hope that individual physicians would and could predictably muster the resources to meet the needs of all those under their care or seeking it. Increased development and use of technology are major factors in steeply driving up the costs of medical services and medical insurance.[41]

1.5 Illustrating the Power of Empathy and the Threats to Its Full Expression

In his recent book, Stanley Reiser has traced the history of technological medicine and its influence on the nature and cost of medical care.[42] When the artificial kidney was invented during World War II, a life-saving device was born that was, in Reiser's words, "the catalyst that forced American society to grapple with the new dilemma of how patients who needed high-technology care could afford access to it."[43] By the 1960s, center-based dialysis was the most advanced and common form of this care.[44] This was the source of the best therapy for chronic kidney disease as demonstrated by a 1971 study comparing the length of survival of those treated: Centers ranked first; dialysis in homes ranked second; and transplants, cadaver organs being the most common source, ranked last.[45] However, the centers required rich resources to meet ever growing demands upon their services. Centers required: An adequate supply of dialyzers; a competent staff of doctors, nurses, social workers, and technicians; administrators able to meet the challenges posed by working at an innovative

---

[41] Daniel Callahan, "Curbing Medical Costs," America 198:8 (March 10, 2008), 9–12. On page 9, he notes that, "health care economists estimate that 40 percent to 50 percent of annual cost increases can be traced to new technologies or the intensified use of old ones."

[42] Stanley Joel Reiser, Technological Medicine: The Changing World of Doctors and Patients (New York: Cambridge University Press, 2009).

[43] Ibid., 31.

[44] Ibid., 42.

[45] Ibid.

frontier; and the funds to pay for all of that. For most individuals, the cost of therapy was beyond their means.[46]

To cope with the costs and demand of these services, a "Medical Advisory Committee" was instituted in 1961 to select among needy patients who would receive treatment at Seattle's dialysis center. Among the criteria employed for these selections were: An age limit set at 17–50 years; financial support; value to the community; a number of a person's characteristics.[47] Before long, this rationing process came to the attention of the media. In 1962, an article appeared in *Life* magazine entitled "They Decide Who Lives, Who Dies," in which the Advisory Committee was called the "Life or Death Committee."[48] The article was skeptical about comparing lives. William Kolff, the physician who pioneered the development and successful use of the machine, being the first to save a life with it, saving the life of a 67 year old patient, admonished the Advisory Committee to "remember that the physician's primary responsibility is the patient."[49] A law review article in 1968 concluded that: "The Pacific Northwest is no place for a Henry David Thoreau with bad kidneys."[50] At this time, other dialysis centers were adopting similar procedures. As Reiser indicates, "The main problem raised by the Seattle selection procedure was its effort to judge the lives of people."[51] However, it was the loss of lives that could be saved that proved to be intolerable for the U.S. public and its government. The empathic emotions were aroused and reflected in the government responses indicated in what follows.

An analysis conducted in 1965 of the number of new patients that required dialysis in the U.S., an estimated 5000–10,000 annually, made it very clear that those who were being treated annually at existing centers (800) were but a fraction of those who would die without such treatment.[52] Equally clear was the fact that financial constraints were creating this large numerical disparity. Furthermore, the plight of those who could not afford care was highlighted in a 1967 article in *Redbook* in which Senator Henry Jackson's alarm over this situation was cited: "It's a terrible thing that probably several thousand men and women will die this year and next who might be saved by an artificial kidney if equipment were available and skilled technicians at hand."[53] The federal government began to respond to what was happening with financial support, and the year 1972 saw approximately 10,000 people on dialysis. In that same year, Congress passed legislation that covered most of the costs of dialysis and transplantation. The legislature declared patients with chronic kidney disease to be disabled and thereby made them eligible for such coverage

---

[46] Ibid.
[47] Ibid., 42–44. A complete list of the Medical Advisory Committee's criteria is on page 44.
[48] Ibid., 44–46.
[49] Ibid., 46.
[50] Ibid., 46–47.
[51] Ibid., 47.
[52] Ibid.
[53] Ibid., 48.

under Medicare.[54] Russell Long chaired the Senate committee that proposed the legislation. In the following reflections, he offers us insight into the societal attitudes and congressional views that led to the passage of such legislation:

> I can recall testimony by a doctor that impressed me. He testified that he had patients with kidney failure—hardworking people, good responsible citizens, honest, salt-of-the-earth people. "What are my possibilities?" they would ask. "Kidney transplant, dialysis, or death," he would reply. "What does it cost?" was their next question. When told, they responded, "There is no way to raise that kind of money. What am I to do?" I sat there and thought to myself: We are the greatest nation on earth, the wealthiest per capita. Are we so hard pressed that we cannot pay for this? A life could be extended 10 to 15 years. You're not going to make any money that way. But it struck me as a case of compelling need.[55]

What we can observe in Reiser's account of the federal government's monetary support for treating individuals with chronic kidney disease is that the same moral imperatives that guide and shape the traditional medical ethic motivated and directed the public and their representatives in government. Once the distress of patients and the loss of lives that could be saved were known and graphically portrayed, the empathic emotions were awakened and expressed as compassion for the distressed and dying, and thus the urge to care for and rescue them from a preventable death. Furthermore, the moral right to what justice demands in the way of medical care was taken on as a moral obligation by society. After all, society is the source of revenue that enabled the government to act as it did. What is more, we should not overlook Medicare itself enacted in 1965. That was a major step toward responding to the increasing inability of physicians and patients to cover the costs of care and lifesaving technology. However, the ever rising costs of medical care that have overwhelmed physicians and patients acting individually, now threaten to overwhelm the ability of governments to assure that the medical needs of all within their jurisdictions are fully and appropriately met even within some very basic level. Having noted that in the last half century society and medicine have sought to apply technological care, Reiser is constrained to say of this effort,

> However as the effort proceeded, the staggering cost of technological medicine surpassed not only the resources of society and its insurance mechanisms to bear. From 1960 to 2006, health care expenditures increased by an average of 2.5 percent per year more than the increase of America's gross domestic product (GDP), which accounts for health care's rising share of GDP. Indeed, from consuming about one of every twenty dollars spent on goods and services…in 1960, health care rose to some one in six dollars by 2006—a path of growth that is unsustainable…studies [show]…that technological change in medical practice [accounts]…for between half to two-thirds of the increase. This is far more than other significant drivers of spending such as increased insurance coverage (from ten to thirteen percent). Technological medicine thus is a paradox—at once a savior and a threat.[56]

This situation should not lead us to disparage or limit the demands of justice, and the strength of our empathic emotions to motivate and guide us to be just. As we have illustrated through some examples discussed in the immediately preceding

---

[54] Ibid., 48–50.
[55] Ibid., 50.
[56] Ibid., 50–51.

paragraphs, the American people will support their government in interventions that extend necessary medical care to those who cannot pay for it. Doing this is in line with the traditional medical ethic. What people will not support are government policies that try to control, curtail, or ration care as the way to reduce costs. That puts the government in conflict with our empathic emotions and thus the natural demands of justice that are integral to our very humanity. Other ways to reduce costs have to be found if justice is to be served. That is the subject to be addressed in Chaps. 4, 5, 6 and 7, and what it would mean to enlist the traditional medical ethic as the norm for just health care policies; and how doing so could lead the way to sustainable medical practices.

1.6 Hippocratic Ethics: A View of Justice Not Found in the Hippocratic Oath

Readers may wonder why our description of the Hippocratic outlook on ethics makes no reference to the Hippocratic Oath. That was done because the Oath expresses an outlook that represents those of a particular school of thought.[57] In the Oath there is no mention of treating everyone and doing so according to their ability to pay even if that means taking no payment at all. Furthermore, the prohibition against doing any surgery is not characteristic of Hippocratic practices: There is ample evidence of resort to surgery by Hippocratic physicians.[58] What is more, the Oath makes no commitment to being scientific in one's orientation and in making clinical judgments. These are among the most significant ways in which the Oath distances itself from the other Hippocratic treatises and the practices of Hippocratic physicians generally.

It is puzzling, therefore, that a recent work by Robert Veatch should claim that what the Oath requires of physicians constitutes the Hippocratic ethic.[59] At most one could think of the Oath as one version of the Hippocratic ethic, albeit a brief one that omits much of the moral tenets of the Hippocratic tradition and stands at odds with important aspects of its practices, most notably surgical interventions. For these reasons, most who agree to abide by the Oath do so in a considerably altered form.[60] In any event, even if one agrees with Veatch that the Oath in its original formulation is not a defensible guide for contemporary physicians, one is not thereby rejecting

---

[57] Ludwig Edelstein, "The Professional Ethics of the Greek Physician," in Ethics in Medicine, Stanley Joel Reiser, Williams J. Curran, and Arthur J. Dyck, eds. (Cambridge, MA: MIT Press, 1977), 40–51.

[58] Ibid., 41. See also Tempkin, Hippocrates in a World of Pagans and Christians, 13, 24–25, and 225.

[59] Robert M. Veatch. Hippocratic, Religious and Secular Medical Ethics: The Points of Conflict (Georgetown University Press, 2012), 27–29.

[60] Ibid., 29 and 68–72.

the key moral commitments that shape the traditional Hippocratic medical ethics and how it understand the demands of justice.

Veatch makes another claim that it is not generally accepted by physicians and should not be. Based on his research, Veatch asserts that the Hippocratic injunction to avoid harming patients is not anywhere in the Hippocratic corpus stated in the exact words of the maxim: "First of all do no harm." Though Veatch may well be correct, the maxim, "First of all do no harm," is widely attributed to Hippocrates. For example, the psychiatrist, Allen Frances, does so in his recent book.[61] He would take issue, therefore, with Veatch's claim that this maxim is not part of the Hippocratic ethic. He finds it in the way Hippocrates practiced medicine.

Frances considers Hippocrates a genius at prognoses, basing these as he did on scientific observations and practical knowledge. Frances lauds Hippocrates for realizing how important it is to sort patients into three groups:

> Those who get better on their own; those who need medical treatment; and those who will not respond to any intervention. This "rule of thirds," the most robust finding in medical history, is still taught to medical students and holds today for many psychiatric disorders.[62]

Basing treatment recommendations on these prognoses, physicians are not to intervene in any way that would cause harm to patients in group one; the same holds for group three so as not to add to the burden of their illness, but instead provide them solace and comfort. Those in group two should receive specific treatments, those that properly balance the risks of treatment over against the risks of illness. Intervention should, on balance, more likely help than harm. Thus, as Frances notes:

> Hippocrates favored caution, mild natural healing except when circumstances required and justified a more aggressive approach. He would be puzzled and deeply saddened by the promiscuous use of harmful antipsychotic medicine that has become a dangerous part of current practice.[63]

It is surely reasonable to view Hippocrates as acting in accord with the intent, if not the articulation, of the admonishment to first of all do no harm. And, as we have discovered from his description of Hippocrates, and the guidance he accepts from him, Frances would be shocked that Veatch would unqualifiedly proclaim: "The Hippocratic ethic is dead."[64] Clearly Veatch is not talking about the ethic that Hippocratic physicians practiced.

Concentrating so exclusively on the Hippocratic Oath as representative of Hippocratic ethics, leads Veatch to mischaracterize not only the moral behavior of the Hippocratics, but also the moral convictions of Hippocrates and those who followed in his footsteps. Veatch, for example, labels as non-Hippocratic the moral

---

[61] Allen Frances, Saving Normal: An Insider's Revolt Against Out-of-Control Psychiatric Diagnosis, DSM-5, Big Pharma, and the Medicalization of Ordinary Life (New York: HarperCollins Publisher, 2013), 241–242.
[62] Ibid., 43.
[63] Ibid.
[64] Veatch, Hippocratic, Religious and Secular Medical Ethics, 159.

commitment to strive for justice in caring for the sick.[65] This assertion flies in the face of what we have previously documented, namely that Hippocratic physicians were praised for their humaneness and their willingness to treat the needs of those they could have overlooked—strangers, aliens, and the poor, including slaves. Indeed, Hippocrates was commended by the celebrated physician Galen for "treating the poor in some small cities rather than joining the court of the Macedonian king."[66]

Hippocratic physicians were decidedly on the side of justice and its pursuit, both in caring for patients, and on behalf of efforts to alleviate poverty. After all, they were willing to treat with equal care those who could pay them nothing and to peg fees for everyone according to their ability to pay. That is the moral commitment to meeting everyone's medical care—<u>universal health care</u>, and doing so equitably.[67]

Veatch does not think medical ethics should be based on codes or rules drawn up by professionals. Medical ethics should draw upon and reflect our naturally shared common morality. However, ironically, that means that professionals, like laypersons, have what it takes to discover how they should behave morally. Hippocrates and his followers certainly did and laypersons expressed their appreciation of their adherence to the demands of justice. Oswei Tempkin cites a prominent fourth century lay orator's accolades for Hippocratic physicians. After elaborately praising them for their love of humanity and their partnership with their patients, he adds this clear portrayal of their actions on behalf of justice in medicine and the larger society, actions he considers to be in accord with their precepts:

> [T]hese precepts have been heeded…to our advantage, by all those physicians who, rather than striving for money, strive for the fame [derived] from having conquered disease. Indeed, I know many physicians who, instead of receiving [money], have themselves spent [money] for poor people. Reasonably so! For the art also affords them respect in the cities. And we look upon the most outstanding of them as if they were gods, believing that in them, next to the gods, lies our hop for deliverance. [68]

For Tempkin, this counts as evidence of what he has been saying about the Hippocratics: "Philanthropy, especially for the poor, regard for reputation, and a sense of equity are all here in this pagan salute to the doctors and their art."[69] One should add, it also documents how rather highly Hippocratics were regarded for their skills, their ability to cure people. Their skill in surgery contributed mightily to their successes and excellent reputation. Tempkin describes a successful operation citing Augustine's account of it in <u>City of God.</u>[70]

Remarkably, Augustine witnessed the surgery so closely that he could describe in detail what was done, how it was done, and what was removed from the patient's body. Even more remarkably, Augustine was privy to the advice the patient was

---

[65] Ibid., 73.
[66] Tempkin, <u>Hippocrates in the World of Pagans and Christians</u>, 220.
[67] Ibid., 220–221.
[68] Ibid. 221.
[69] Ibid.
[70] Ibid., 223–225.

given by physicians, advice that the patient at first rejected vehemently. However, the patient's suffering and the opinions of several physicians that the surgery was needed to relieve it, finally persuaded the patient to undergo the operation. Augustine was in a position to agree or disagree with the arguments of the physicians: He chose to agree and joined the patient in praying for success.

In seeking the consent of their patients and in providing their knowledge-based rationale for that consent to their services, these Hippocratic physicians were acting in accord with clear admonitions found in the Hippocratic treatise <u>Decorum</u>. Physicians there are exhorted to posses the skills to make the case for any of their interventions; they are to present proof that what they recommend is the correct course to follow.[71] Tempkin provides the reason why being skilled in argumentation was so important.

> The great rhetorical skill demanded of the physician was a desideratum of medical practice because opinions had to be justified to the patient and his relatives and friends, and defended against objecting bystanders and colleagues.[72]

For such justification, Hippocratics, like contemporary physicians, relied upon the best empirical data they had available at the time.

It is quite startling, therefore, to read Veatch's unqualified belief that Hippocratic physicians did not observe the requirement of obtaining informed consent![73] Of course, Veatch limits what he regards as the Hippocratic ethic to the Hippocratic Oath as he interprets it, and draws inferences from it.

What we have sought to accomplish in this brief excursus prompted by Veatch's recent publication, is to reject his claim that the Hippocratic ethic is to be found in the original Hippocratic Oath. This Oath, a sectarian minority document, does not reflect the Hippocratic ethic we find in the other Hippocratic treatises and in the practices of Hippocrates and his Hippocratically educated followers in the ancient world. That is the ethic we are invoking.

But why invoke this ancient ethic? We do so because we are never going to achieve justice with respect to who receives medical care and what care they receive if we do not accept and implement some key demands of justice as articulated, taught, and acted upon by Hippocrates and the main body of Hippocratic physicians.

What are these key demands of justice? The overarching one is that everyone ought to be able to obtain what medical care they need regardless of whether they can pay for it fully or not at all. A second key demand of justice is that of putting a high premium on the avoidance of harmful interventions, especially those that are unnecessary or otherwise scientifically unjustifiable. Thirdly, justice demands the commitment to determining which therapies are most efficacious and least harmful, and which are to be rejected outright, and making these determination s on a sound scientific basis. In Chaps. 4, 5, 6 and 7, we will indicate how the implementation of

---

[71] Ibid. 26.
[72] Ibid., 27.
[73] Veatch, Hippocratic, <u>Religious and Secular Medical Ethics</u>, 14.

these elements of Hippocratic justice markedly increase the quality of medical care while at the same time greatly reducing costs so much so that universal healthcare and achieving justice becomes a plausibly attainable goal. As we will also document in Chap. 4, there are growing efforts by physicians in the various medical specialty organizations to reduce unnecessary and scientifically unsupported interventions, thereby reducing harm to patients and simultaneously reducing significantly the costs of practicing medicine.

Now, however, we turn to Chap. 2 to delineate cognitive processes that serve to yield a way of justifying the demands of justice as we have spelled them out in the present chapter.

Chapter 2
The Cognitive Bases for Deciding When Policies Are Just

Abstract It has long been recognized that, psychologically, we experience *ought*, a moral demand, when we relate a choice before us to our conception of an ideal self. Based on this reality, ideal observer theories have been promulgated for centuries. I draw on Roderick Firth's most precise version. What is right is defined by the approval of an Ideal Observer who is omniscient with respect to factual knowledge, omni-percipient in imagining how everyone is affected by one's decision, and is impartial. Drawing on neuro-ethical research and other psychological studies, I support these criteria for moral decision-making and argue for changing how impartiality is understood. Impartiality is not attained by being disinterested and dispassionate but rather by having equal empathy for everyone, including oneself, as in the Golden Rule.

In Chap. 1, we identified some of the fundamental moral responsibilities, and the rights correlatively related to them, that comprise the moral imperatives of justice. We say "some" because we are not claiming to have provided an exhaustive list of moral responsibilities and rights that may properly qualify as universally existing moral requisites of individual and communal life. By having recourse to justice in its generic sense, that is, as what we owe one another as human beings, we could uncover and defend the Hippocratic ethic, and the moral imperatives of justice that guided those whose practices were in accord with its admonishments. We noted and discussed the following central tenets of the Hippocratic ethic: Physicians should treat their patients, however rich or poor, on the basis of their medically diagnosed needs alone; they should first of all do no harm; they should offer all who seek their help only those interventions that are necessary and in a patient's best interest; and, in accord with their skills, and empirical knowledge, they should provide or seek such care for anyone they have determined to be in need of it.

Although there are physicians who do everything in their power to practice medicine in conformity with the traditional medical ethic, the present structure of the medical complex and the high, ever increasing costs of the care being delivered, are formidable barriers to doing this successfully. Thus we find that the medical insurance plans available to meet these high costs are very costly, and yet do not cover costs completely, nor do they cover or pay for every diagnosed need. Similarly, the U.S. federal and state governments do not pay completely for the medical care provided, nor for all medical needs, through their Medicare and Medicade programs, laudable though they are in many respects and much needed as a safety net for many. At the same time, among those who advocate universal access to medical care as a matter of justice, the dominant view is that the governments should guarantee minimum, basic medical care, a policy that falls short of providing the medical care the traditional medical ethic calls for as a matter of justice. We will examine and argue against this view in the next chapter.

And so we find that the federal government and insurance companies in the United States do not provide for equality in the amount and kind of care people can receive given their economic status, nor do major theories of justice that we examine in Chap. 3 demand that governments anywhere guarantee such equality. Of course, the constantly rising costs of medical care would seem to make it beyond the reach of governments and insurance companies to assure that everyone is financially able to receive all the medical care they need.

In Chaps. 4 and 5, we will directly address how it is possible to implement a policy that is in accord with the Hippocratic ethic. For that purpose, it will be necessary to demonstrate that there are just ways greatly to reduce the costs of medical care. In this chapter, our concern is to provide a justification for insisting that full medical care for everyone is what justice demands. This we will do by spelling out the cognitive processes that serve to justify a policy that we believe to be just, and by doing so, provide criteria by means of which to guide policy decision-making and deal with disputes.

2.1 Moral Decision-Making: Identifying Its Cognitive Character

Regarding any policy we are following or deciding to undertake, we should consider whether it is morally right or wrong. What do we mean when we refer to some activity or set of activities as a policy? A policy is a designated goal or set of goals, and an action or set of actions, designated as the means to achieve that goal or those goals. Policies are almost always morally complex. Any decision about whether any given policy is just will be a decision about what policy, given the alternative, is the most morally justifiable one to adopt. The reason for that is due to the various right and/or wrong-making characteristics of the action or actions the policy adopts as a means to achieve its goal or goals. Deliberations as to what policy can be considered just, and more just than other possible policies, will involve a process: of identifying all the right and wrong-making characteristics of the policy's goal or goals and of the activities the policy engenders; of determining the relative weight to give to each of these characteristics; and of resolving any conflicts that may exist among them. The process we are describing is that of making a moral decision, and what we want to know about moral decision-making is the cognitive basis we have and employ for making them, and for justifying what we have decided.

Many of our experiences are experiences of what ought to be the case or true of reality. Such experiences are not exclusively moral experiences. What ought to be true of reality is an experience that occurs in at least three cognitively differing ways: In sensory observations; in logical reasoning; and in empathically driven proclivities and inhibitions. Hence, we seek knowledge of reality and verify our discoveries by means of our five senses, reason, and our empathic emotions. As we will indicate, we make use of all of these cognitive abilities when making a moral decision that a particular act or policy is just, and one we ought to continue or implement.

To begin with, consider the process involved in obtaining knowledge by means of sensory observations. Take the example of keeping track of a child's growth in height as one of the indications of whether that child is properly nourished and healthy. Against the background of previous measuring of the growth of many children, the period and range of growth typical of healthy children has been established as a norm. On this basis, one can say that a child of a certain age ought to be growing and achieving a height that falls somewhere within this normal pattern.

The parents of Sally bring her to a physician for a physical examination. To check her height, the physician uses a device marked in inches and feet to see with his eyes how many feet and inches Sally measures from the top of her head to the bottom of her feet. As it turns out, the physician is pleased to report to her equally pleased parents that Sally is four and a half feet tall, a height that is average for her age and a reasonable increase since she was last measured. Since other measures of her condition also turn out favorably, Sally can be said to be thriving. The fact that Sally is four and a half feet tall, and the fact that she ought to be approximately that height if she is to be judged to be growing normally for her age, are non-moral facts. They simply constitute pieces of information obtained by sensory observations of her and many children before her time. However, the fact that the parents and physicians ought to be checking whether and how much Sally is growing is a moral fact. They are concerned with nurturing this child, making as certain as they can that she is receiving the necessary care needed to protect her health, avoid suffering, and protect her very life. Shortly we will examine the cognitive basis they have for deciding that they ought to do what they are doing on behalf of Sally.

By using our senses, we can know how tall Sally is. Using a process of logical reasoning we can know something more about Sally's height. We can find out by reason alone how many yards tall she is. If, for example, we know that twelve inches is in a foot, we can calculate that four and a half feet equals four times twelve plus six inches or a total of fifty-four inches. If we have learned that there are thirty-six inches in a yard, we can declare that fifty-four inches equals thirty-six plus eighteen inches or a yard and a half. At the proper grade level, a teacher could use this example as a test of her pupils' mathematical knowledge and reasoning skills. That four and a half feet ought to be the equivalent of one and a half yards is a non-moral fact of mathematical logic.

How does the experience of a moral ought differ from what ought to be the case with respect to Sally's height as measured in feet and as calculated mathematically in yards? The psychologist Gordon Allport describes the experience of what ought to be the case morally as follows: "I experience 'ought' whenever I pause to relate a choice that lies before me to my ideal self-image."[1] The height of Sally is not a matter of choice, nor does it depend upon one's self-image: It simply is as measured and as deduced. However, whether her parents and their physician ought to keep track of Sally's growth in height to check the nurture she is receiving and how she is responding to it, is a choice that does relate to the ideal self-image of her parents and their physician as being compassionate and conscientious in providing for Sally's health, life, and well-being.

Although the height of Sally as such is not a moral fact and the evidence for it is visual, accurate measurements of height, and truthfully reporting these, are moral actions. Whether we take every precaution accurately to ascertain what is true and truthfully to convey what we have discovered, is a matter of the kind of person we *ideally* aspire to be and to have others see us as being.

Conscientiously seeking to discover what is true, and faithfully sharing what is true with others, are behaviors essential to the very possibility of accumulating and teaching knowledge. The whole scientific enterprise would be impossible if people did not universally have a proclivity to be truthful and to be inhibited about lying. As we noted in Chap. 1, our individual and communal lives depend upon our widespread compliance with and, as necessary, enforcement of, these proclivities and inhibitions. Sally's well-being, and proper nurture to that end by her parents, require such morally trustworthy behavior from their physician. The parents expect it and rely upon it; they would be loath to retain a physician they did not trust to be careful, caring, and truthful. Being truthful should include honest representations of the physician's own expertise; of the special expertise available from any array of other physicians, and of the opportunities and limitations of the art of medicine more generally. Being such a physician is being an ideal self, and in the ways noted, an ideal physician.

Is there any scientific evidence that we as human beings are disposed to achieve accuracy with respect to what we perceive and disposed truthfully to answer any inquiries about what we saw? Furthermore, is there any scientific evidence that we generally choose to do so, though under some circumstances we choose not to do so, and that, in both instances we refer such choices to our ideal self? The social psychologist Solomon Asch conducted a controlled experiment that shed considerable light on these very questions.[2]

In Asch's experiment, the task of subjects, in a series of trials, was to identify which two lines of four presented were equal in length. The task was clear enough perceptually that subjects, individually presented with these lines, reported the correct answers 92.6% of the time: They didn't simply see correctly and almost per-

---

[1] Gordon Allport, Becoming (New Haven: Yale University Press, 1955), 13.
[2] Solomon Asch, Social Psychology (Englewood Cliff, N.J.: Prentice Hall, 1952).

fectly so; they faithfully reported what they had seen. They were definitely prone to tell the truth.

Asch then tested to find out what would happen when individuals are subjected to peer pressure. He gave the same series of trials in the presence of seven to nine persons of the same age group (college students), instructing them to give some wrong answers and to do so unanimously. Now the naïve subjects gave correct answer 66.7% of the time rather than the 92.6% of correct answers previously offered by individuals making these judgments individually on their own. Indeed, 58% of the naïve subjects who were confronted by the group instructed to give certain number of incorrect answers were definitely influenced to offer as correct some or all of these same erroneous responses.

When Asch interviewed the subjects of these trials, each was aware of the contradiction between what each at the outset regarded as correct perceptions and reports of what was perceived, and what the instructed group was sometimes unanimously reporting. These subjects were also aware that what was at issue were perceptions of fact, yielding only one result that was true to the facts despite the disagreements arising from the group instructed to give some answers that were discernibly wrong. Although they were perplexed and even doubtful about themselves in this situation, those subjects who chose to be true to what they saw, and faithful in conveying what they saw, explicitly told Asch they were determined to behave exactly as they did: They were pleased to learn that their responses were correct and pleased that they had been true to their convictions. Some of the subjects who had been influenced to give the same mistaken and false answers given by the instructed group felt some guilt or regret once apprised of what had been happening. They were not indifferent about what kind of persons the experimenter thought they were.

This experiment does provide some evidence that our ideal image of ourselves is of one who conscientiously strives to discern reality accurately and convey what we discover truthfully. And, under less social pressure to doubt our abilities, or be seen as out of step with the behavior of a distinct majority of our peers, we are definitely prone to be accurate and truthful. Asch tested this by giving some subjects the same experimental task in the presence of one other subject instructed to offer the same erroneous responses offered by the groups of seven to nine in the previous experiment. Not one of these individual naïve subjects gave erroneous answers; they neither made errors nor reported any. They did this despite their considerable astonishment about the errors the other person was making. Clearly they expected others to be like them and to act as they did, particularly in a situation in which it was so easy to be accurate and truthful.

The expectation that others will or should have very same experience of what ought to be done morally in similar circumstances indicates that the ideal self, the self we aspire to be, is a shared cognitive perspective that is characteristic of us as human beings. Hence, the moral ought is not experienced as a merely personal or subjective impulse or desire to act in a certain way. The psychologist Fritz Heider has expressed this very clearly as follows: The perception of a moral ought is the perception of a "vector which is like a wish or a demand or a requirement on the part of some suprapersonal order and which has the validity of objective existence."[3] Such a suprapersonal order is objective in the sense that its demands have interpersonal validity and people generally concur in its demands.

Heider is well aware that when we perceive that someone or some group wants us to act in a certain way, we recognize a vector in our environment and an objectively existent wish or demand. However, unless the wish or demand is perceived to be the wish or demand of the whole objective order, it will not be perceived as a moral demand. That moral demands are demands of the whole objective order, is a phenomenon that has been empirically observed.

Consider the observations, made by the psychiatrist Robert Jay Lifton, of Vietnam veterans who were profoundly disturbed and overcome with guilt over their involvement in situations in which innocent civilians were killed, including children. As a survivor of My Lai portrayed it:

> Here in Vietnam they're actually shooting people for no reason...Any other time you think, it's such an extreme. Here you can...shoot for nothing...it's even smiled upon.
> ...Good for you. Everything is backwards. That's part of the unreality of the thing... Something [at My Lai] was missing...that made it seem like it wasn't happening.[4]

Lifton distinguished two recurring images present in the struggle of these veterans with their guilt, some over having killed innocent persons, some over having been unable to stop such killing: The first was that their transgressions caused "a wound in the order of being"; the second was "an image of a world beyond the transgression itself." What this appeal to a world beyond meant to these veterans is that if one is "to transcend the condition of the transgression (the atrocity-producing situation) one had to open oneself up to the larger 'order of being' one had injured."[5] In fact, these veterans spoke of psychiatrists and chaplains who sought to assuage their guilt over the killing of innocent civilians—including children—as representing to them a "counter universe," the title of Chap. 6 of Lifton's book. They had an intense anger for those chaplains and psychiatrists who "blessed" what they called murder and atrocities. Expressing how these veterans perceived these chaplains and psychiatrists in the language of Heider, they saw them as not at all open to the demands of the "whole objective order," blind therefore to what ought to be the case morally.

Confronted by their guilt, these veterans are now, in retrospect, comparing the self they were when engaged in, or complicit in, killing innocent individuals, with their ideal self, that is the self they ideally wish to be and wish they had been: Ideally, they wish to be someone who is totally inhibited from committing what amount to murderous actions, and someone who is moved to try to prevent such actions on the part of others. By being capable of feeling guilty about their past moral failures, and by being willing to deal with their guilt, they are beginning to go

---

[3] Fritz Heider, The Psychology of Interpersonal Relations (New York: John Wiley and Sons, 1958), 219.
[4] Robert Jay Lifton, Home from the War (New York: Simon and Schuster, 1973), 36–37.
[5] Ibid., 103–104.

beyond their presently clear acknowledgements of how they ought to have behaved, to a consideration of what it is about themselves that has led them to perceive clearly how one ought to behave morally, and to increase their confidence that their moral perceptions and judgments are correct, and that those who differ with them are definitely mistaken, blind to the objective demands of morality.

There are numerous references in the philosophical literature to an ideal self serving as a guide in our moral deliberations over what we ought or ought not to do. Kant speaks of the business of conscience transacted "as if at the command of another person...this other might be an actual or merely ideal person which reason frames to itself."[6] Adam Smith asserts that we have an actual disposition to judge our actions with "the eyes of other people or as other people are likely to view them....We endeavor to examine our own conduct as we imagine any fair and impartial spectator would examine it."[7] Westermark observes that: "Almost inseparable from the judgment we pass on our own conduct seems to be the image of an impartial outsider who acts as our judge."[8] G. H. Mead points to a similar phenomenon using the notion of a "generalized other."[9] The theologian, H. R. Niebuhr, after citing each of these scholars, describes how we are cognitively guided to choose morally responsible actions, namely by interpreting our notions "as taking place in a universe," and also by "understanding that there will be a response to our actions by representatives of universal community, or by the generalized other who is universal, or by an impartial spectator who regards our actions from a universal point of view, whose impartiality is that of loyalty to a universal cause."[10]

These scholars are portraying our deliberations as to whether some actions we might undertake or have undertaken, as an effort to adopt the impartial perspective we imagine to be characteristic of ideal moral spectators or judges of the action in question. The late Roderick Firth refined and expanded this way of thinking about the cognitive processes by means of which we make moral decisions and justify them.

2.2 The Cognitive Processes of Justifying Moral Decision

Firth understood impartiality as a characteristic of an ideal observer, a hypothetical being serving as a device to delineate what characteristics anyone would have to possess to "reacting in a manner which will determine by definition whether an ethical judgment is true or false."[11] According to Firth, "X is right" means "X would

---

[6] H. Richard Niebuhr, The Responsible Self (New York: Harper & Row, 1963), 74.

[7] Ibid., 75.

[8] Ibid., 76.

[9] Ibid., 76–77.

[10] Ibid., 87–88.

[11] Roderick Firth, "Ethical Asbolutism and the Ideal Observer," Philosophy and Phenomenological Research 12 (1952), 321.

be approved by an ideal observer who is omniscient, omnipercipient, disinterested, and dispassionate, and otherwise a normal human being."[12] By means of these characteristics, Firth identified certain cognitive conditions, the approximation of which, enter into our deliberations as to whether what we experience as a moral demand is truly a moral demand. We can only approximate what is ideal because, as Firth indicates, the term "ideal" means that an ideal observer is "conceivable" and has "certain characteristics to an extreme degree" in the same sense in which "we speak of a perfect vacuum or a frictionless machine as ideal things."[13]

What grounds are there for selecting omniscience, omnipercipience, and impartiality to characterize the ideal cognitive conditions for guiding our assessments and justifications of our moral perceptions and judgments? For such grounds, Firth asks us to consider the actual cognitive processes we would use to resolve doubts, or conflicts with others, about what we perceive or judge to be morally right or wrong, as the case may be. That is how we shall proceed. It should be noted that our account of impartiality will differ from Firth's account of it as disinterestedness and dispassionateness.

2.2.1 Moral Cognition: Knowledge of Non-moral Facts

We are prone to refrain from making moral judgments about some actions or policy if we are convinced that we lack sufficient knowledge of non-moral facts to do so justifiably. Indeed, "we regard one person as a better moral judge than another if, other things being equal, the one has a larger amount of relevant factual knowledge than the other."[14] Thus, for example, primary care physicians will refer their patients to specialists when more knowledge of the cases in question is deemed essential to securing a more predictably good outcome for those patients by means of obtaining correct diagnoses, and/or the best treatments currently available. Furthermore, when moral disagreements exist, one avenue we have for trying to resolve such disagreements is to make certain that the persons who disagree are equally aware of the relevant non-moral facts, and equally free of false beliefs about any such facts. Granted that this is a process we consider apt, it raises the question as to how much knowledge of non-moral facts we need to provide the parties to a moral dispute to assure that they know all the relevant non-moral facts and are free of false beliefs regarding them. The answer to that question, simply stated, is nothing less than complete knowledge of the non-moral facts; nothing less, that is, than omniscience with respect to them.

Firth was aware that some would object to invoking omniscience as an ideal cognitive condition for making moral decisions. Why not limit what has to be

---

[12] Ibid.
[13] Ibid.
[14] Ibid., 333.

known of the non-moral facts to those that are relevant? Because, as Firth has pointed out:

> We apparently believe not only that the "facts of the case" are relevant to the objective rightness or wrongness of a particular act, but also that there is no point at which we could be logically certain that further information about matters of fact (e.g. further information about the consequences of the act), would be irrelevant. A satisfactory ethical analysis must be so formulated, therefore, that no facts are irrelevant *by definition* to the rightness or wrongness of any particular act.[15]

In other words, whether a particular non-moral fact is relevant or irrelevant to making the right moral decision in a given instance requires us to know that fact. We can only know how many of such facts are relevant once we know and consider all of them. Ideally, that demands omniscience with respect to the non-moral facts.

Since omniscience is an unattainable ideal, it means that some uncertainty attaches to our moral judgments. That, however, is true of all our knowledge, even of the hard sciences. At one time, the atom was characterized as the smallest bit of matter. Now, with more facts in hand, that is no longer considered to be true. What we know to be true, based on our cognitive standards for verifying the claims we make, is nevertheless always subject to revision as additional facts are discovered, and our ability to verify what we claim to know is further refined and/or expanded. Surely we should not expect the situation for making moral judgments to be any different. Indeed, the ideal omniscience demands us to be open to necessary changes in our moral judgments and to seeking increased understanding as our knowledge of non-moral facts increases.

Pragmatically, then, the ideal of omniscience sets a high bar for making our moral decisions about which actions or what policies are the most just ones to adopt and set in motion: We are to be exceedingly conscientious in our pursuit of knowledge about all the non-moral facts that could possibly be relevant to making those decisions. At the same time, attaining decisions for which we can offer plausible justifications, do, and logically must, invoke additional cognitive processes. The very effort to gather sufficient and accurate non-moral facts will be thwarted or distorted if we fail to be able vividly to imagine how people will be affected by the actions or policies under consideration, and if we approach our deliberations with partiality rather than with an impartial perspective. These aspects of moral cognition we will next consider.

---

[15] Ibid., 334.

2.2.2 Moral Cognition: Vividly Imagining How People Are Affected by Our Actions or Policies

We sometimes disqualify ourselves to make moral judgments about certain actions or policies because we are unable sufficiently to imagine or visualize some of the relevant facts to be taken into account. In fact, "we regard one person as a better moral judge than another if, other things being equal, the one is better able to imagine or visualize the relevant facts."[16] The influence of our ability to visualize the consequences of our actions or inactions is well understood. An appeal in the mail to help starving children is known to be more likely to elicit a positive response if these children, or at least one of them, is graphically depicted. Even more likely to have us decide to offer assistance to a starving child is to have that child appear at our very front door. Confronted in this way by the very visible effects of starvation, we can easily imagine the dire consequences of failing to provide, or seek what it would take, to prevent the further suffering and possible death of the child we see before us: Our empathic emotions are forcefully triggered by what we envisage.

As we noted in Chap. 1, empathy functions cognitively to enable us to know how someone else feels. Neuroscientists speak of that as empathic accuracy. Of ancient origin, and yet still advocated and serving as a cognitive guide for making moral decisions, the Golden Rule presupposes that we have this capacity for empathy, and that we can best vividly and accurately imagine how given actions or policies will affect others if we imagine how they would affect us. The Golden Rule is a cross-cultural phenomenon; it is framed in various world religions and is formulated in the following ways:

Baha'i: "And if thine eyes be turned towards justice, choose thou for thy neighbor that which thou choosest for thyself." Lawh'i 'Ibn'i Dhib, "Epistle to the Son of the Wolf" 30

Buddhism: "Hurt not others in ways you yourself would fnd hurtful." Udana-Varga, 5:18

Christianity: "In everything do to others as you would have them do to you; for this is the law and the prophets." Matthew 7:12

Confucianism: "Do not unto others what you do not want them to do to you." Analects 15:13

Hinduism: "This is the sum of duty: do naught unto others which would cause you pain if done to you." The Mahabharata, 5:1517

Islam: "Not one of you is a believer until he loves for his brother what he loves for himself." Fortieth Hadith of an-Nawawi, 13

Jainism: "A man should wander about treating all creatures as he himself would be treated." Surtrakritanga, 1:11:33

Judaism: "What is hateful to you, do not do to your neighbor: that is the whole of the Torah; all the rest of it is commentary." Talmud, Shabbat, 31a

Native American: "Respect for all life is the foundation." The Great Law of Peace

Sikhism: "Don't create enmity with anyone as God is within everyone." Guru Granth Sahib, page 259

Taoism: "Regard your neighbor's gain as your own gain and your neighbor's loss as your own loss." T'ai Shang Kan Ying P'ien

---

[16] Ibid., 335.

Zoroastrianism: "That nature alone is good which refrains from doing unto another whatsoever is not good for itself." Dadistan-I-Dinik, 94:5[17]

All of these formulations of the Golden Rule reflect the concern all of these religions share, that their adherents be able to know the difference between right and wrong, act on behalf of what is right, and avoid what is wrong. It is an admonition, formulated by design to engage and heighten one's empathic emotions, empathic accuracy to be sure, but also sympathy and compassion. The existence of the Golden rule, then, certainly attests to the longstanding and widespread recognition, that vividly imagining how others are affected by our actions and policies, is one of the essential cognitive conditions for making correct moral decisions.

By itself, the Golden Rule as cited, is a practical device to assist individuals to approximate the ideal conditions for obtaining moral knowledge. As Firth has indicated, these ideal conditions include nothing less than vividly imagining "all actual facts and the consequences of all possible acts in any given situation." He refers to this as "omnipercipience" or "universal imagination" that guarantee that "ethically-significant reactions are forcefully and equitably stimulated."[18] Note the expression "equitably stimulated." That means that the Golden Rule should function impartiality. We turn then to consider the meaning and role of impartiality in moral cognition.

2.2.3 Moral Cognition: Impartiality

According to Firth, "we sometimes disqualify ourselves as judges of certain ethical questions on the ground that we cannot make ourselves impartial, and we regard one person as a better moral judge than another if, other things being equal, the one is more impartial than the other."[19] In fact, judges who preside in our courts disqualify themselves, that is, recuse themselves from a case whenever they are convinced or even suspect that they cannot be impartial with respect to that case. We expect of judges, and judges expect of themselves, that in cases they are deciding over which they have jurisdiction, they will deal with all the facts and conditions presented them objectively, that is without distortions by reason of personal interests or feelings. Thus, for example, judges would feel compelled, or be urged by others, to recuse themselves from a case involving a company in which they own shares, a case involving a personal interest. The same demand for recusal applies to cases in which judges would be influenced by their personal feelings, such as cases in which the victim or the defendant is a family member or close friend.

In one of its commonly acknowledged senses, then, impartiality is what we attribute to someone who is objective, whose perception and understanding of facts or conditions are free of distortions that can result from personal interests and feelings.

---

[17] These are compiled by the Tannebaum Center for Interreligious Understanding.
[18] Ibid.
[19] Ibid.

The path to impartiality, therefore, would appear to be a cognitive perspective that is uninfluenced by personal interests and feelings. Firth expresses this very widespread understanding of impartiality when he speaks of it as a state of being disinterested and dispassionate. For Firth that means impartiality can only be achieved by an ideal observer who lacks any particular interests, and is incapable of experiencing any emotions that are directed toward any objects or persons because they are thought to have one or more essentially particular properties, emotions such as jealousy, self-love, and personal hatred.[20] However, to be without any particular interests and emotions, especially self-love, would thwart the achievement of impartiality. How can that be?

Consider an example of how impartiality is sought, drawn from the practice of medicine. Suppose Frank, a surgeon, and other physicians he has consulted, have determined that a child of his requires lifesaving surgery. Suppose also that Frank has the requisite skills to perform this operation and would do so if no one else was available to do so within the time required. However, the ideal in this situation is to request the services of another physician who is no less skilled in this particular kind of surgery. Why would Frank prefer to recuse himself from operating on his own child? Precisely because of his intense interest in a successful outcome. It is best for his child to have such a critical operation performed by a surgeon who does not have Frank's heightened inhibition against inflicting harm on his own greatly beloved offspring, and who does not have Frank's fear of failure, and of being the one possibly responsible for failure, should that be the outcome. Frank wants for his child a physician as skilled as he is but with the steady composure typical of a highly successful surgeon, emotionally prepared to subject patients to the pain and risks of surgery for the sake of saving a life and restoring health. It is wise and compassionate for Frank not to trust himself to operate on his own child, especially when the stakes are high. He recognizes that his intellectual and physical abilities are at risk of being compromised by the sheer magnitude of his fear of failure and aversion to inflicting pain on his own child. In his regular practice, Frank confidently overcomes such fears and aversions. He achieves an impartial perspective, that is, he treats his patients as he himself would wish to be treated were he to be similarly in need of the services he is rendering. It is a physician who can be similarly impartial toward his child that Frank realizes is called for.

Note that we are describing impartiality as a matter of overcoming certain particular, strong emotions rather than being devoid of them. Indeed, Frank chooses Carol, a highly accomplished surgeon, who will treat his child as she would hope her own child would be treated. In short, Frank has good reasons to choose an empathic physician on behalf of his child. The difference between Frank and Carol is not that Carol is free of empathic emotions directed towards Frank's child; rather Carol's positive regard for Frank's child is not nearly as intense as Frank's. She has no history of invested hope in that child's future, and she will not have to live with memories of past happy days together and thoughts of what might have been in the

---

[20] See Ibid., 335–341, for Firth's detailed definitions of particular interests and emotions and his carefully crafted arguments for his view of what constitutes impartiality.

2.2 The Cognitive Processes of Justifying Moral Decision

event that the operation fails to save the child's life. She is prepared to do for Frank's child what she has done for many other children. As a successful surgeon, she has demonstrated her ability to overcome her empathically driven inhibitions against inflicting pain and her empathically driven fear of failure. She has proven to be able to do this for all her patients and in that respect has achieved the level of impartiality needed to deter the influences of overly heightened particular emotions and interests that would otherwise undermine her rational and objectively oriented competencies as a surgeon.

What this hypothetical, though realistic example illustrates is that, in a case like Frank's, the path to attaining impartiality is guided by, not thwarted by, our particular empathic emotions, even our own most intense particular interests in, and attachment to a child of our own. In this case, our experiences of these emotions, known to be a distorting influence, combined with our heightened concern for a child of our own, leads us to seek and secure the conditions that avoid distortions of impartiality. Self-love is also involved. It is empathy for himself that moves Frank to avoid the deep sorrow he would feel if he caused a failed attempt to save his child's life. Empathy for himself and for his child leads him to bend every effort to avoid this outcome in the manner we have discussed.

Our depiction of impartiality has so far been describing it insofar as it pertains to overcoming expressions of emotions that undermine objectivity and rationality. Impartiality also refers to inclusiveness, to overcoming the partiality associated with excluding some person, or group of persons, from the circles easily encompassed by our empathic emotions. We find ample evidence for the existence of efforts to be impartial by being inclusive. There are regulations and laws that are aimed at preventing discrimination against individuals and groups: Individuals and groups are not to be denied justice by reason of their nationality, race, ethnicity, class, religious affiliation, gender, or sexual orientation. If impartiality is to be served, every human being should be treated as a member of the moral community, the community in which its members offer one another, and receive from one another, the benefits of the universal responsibilities and rights that are requisite for the very existence, sustenance, and protection of individual and communal life. This is the impartial perspective that underlies and undergirds our humane treatment of those who are criminals or our enemies in war. When these are our prisoners, for example, we are to provide what is necessary to meet their physical needs, including medical care, and refrain from any cruel behavior such as what is widely regarded as torture. In our hospital emergency rooms, all in need of care are attended to without regard as to who they are or what behavior of theirs or others brought them there.

Amartya Sen deals very perceptively with inclusiveness in the sense described immediately above, but also with inclusiveness as a way of expanding our cognitive ability to be impartial. He does this by developing a conception of what he calls "open impartiality," reflecting a viewpoint he discovers in Adam Smith's portrayal of an impartial spectator.[21] His idea is that since all of our perspectives are limited,

---

[21] Amartya Sen, The Idea of Justice (Cambridge, M.A.: Harvard University Press, 2009), 124–152. Sen also discusses the relationship between impartiality and objectivity. See ibid., 114–123.

an impartial perspective is cognitively best approximated by dialogue that is inclusive of a diversity of perspectives. Deliberations and decisions regarding what is just or unjust, therefore, should be inclusive, open as much as possible, to all kinds of people and points of view. For that reasons, Sen is critical of John Rawls for having people choose principles of justice from the perspective of "closed impartiality," that is, from the perspective of members of a particular society or political entity.[22]

The concern for practicing open impartiality and resisting closed impartiality has a long history. One can find this concern in discussions of what we owe our neighbors. As Sen notes, "the question of one's duty to one's neighbor has a huge place in the history of ethical ideas in the world."[23] The notion that the neighbor is limited to one's proximate neighbors or any group of people to whom one feels loyalty, such as a particular nation, persists alongside long standing resistance to this notion of what Sen refers to as a "fixed neighborhood."

Sen considers the case for fixed neighborhoods to be intellectually fragile, and he asserts that this is a point,

> made with compelling clarity by Jesus of Nazareth in his recounting of the story of "the good Samaritan" in the Gospel of Luke [10:25–32]. Jesus's questioning fixed neighborhoods has sometimes been ignored in seeing the good Samaritan story as a moral for universal concern, which is also fair enough, but the main point of the story told by Jesus is a reasoned rejection of the idea of fixed neighborhood.[24]

This argument against a fixed neighborhood comes in response to the direct question about who qualifies as one's neighbor posed by a local lawyer after Jesus has commended him for affirming that he has the obligation to love his neighbors as he loves himself. Jesus tells him the story of a severely wounded man who was helped by a traveling Samaritan, though not by a priest or Levite, both of whom offered no assistance, choosing rather to cross over to the other side of the road and walk by.

In the light of these events, Jesus asks the lawyer who it was that turned out to be a neighbor to the wounded man. The lawyer cannot in good faith avoid identifying the Samaritan as the one who proved to be the neighbor, the one who showed compassion on the wounded man by dressing his wounds and paying for further care at an inn. That identifies as a neighbor to this lawyer one who lives some distance from his proximate neighborhood and who belongs to another nation, the members of which are generally disliked by the lawyer's fellow Jews. So much for fixed neighborhoods. Being humane and practicing humaneness is universal, not limited to any one group or people. Jesus confronted the lawyer with this reality. The contemporary neurosciences confront all of us with this same reality: The empathic emotions, which include compassion, have their source in the human brain.

In this exchange with the lawyer, Jesus endorses an inclusive version of the Golden Rule. Loving your neighbor as yourself means you should treat your neigh-

---

[22] Ibid., 123–152.
[23] Ibid., 171.
[24] Ibid.

bor as you treat yourself: Be as empathic toward your neighbor as you are toward yourself. When Jesus advocates following the Golden Rule stating, "In everything do to others as you would have them do to you" (Matthew 7:12, NRSV), those "others" are your neighbors and your neighbors include all human beings. As history attests, achieving this kind of open impartiality has been a constant struggle, to say the least for individuals and groups of all kinds including nations. That struggle continues.

We learn something else from the exchange between Jesus and the lawyer. Without self-love, that is, empathic emotions directed toward ourselves, the Golden Rule will not serve as a cognitive source of knowing how to behave in a morally proper way toward others. If we were to hate ourselves and thus be willing to harm ourselves, we would have a basis for inflicting harm on others or inviting harm from them. That is what we find in the killing of twelve students and a teacher at Columbine High School in 1999; the two students who did this afterwards committed suicide. In fact, in the six similar incidents in the U.S. of individuals killing five or more innocent people, occurring between 1999 and 2007, five of the perpetrators committed suicide before they could be stopped or apprehended by others. In the other case, the killer persisted in putting himself in harm's way, shooting his victims in the presence of the police until he was predictably killed by those very officers sworn to defend innocent lives.

The Golden Rule, as stated in the Gospel of Matthew, is portrayed as a guide to those who exert empathy on behalf of life. Jesus uses the example of parental love to indicate what is required to make the Golden Rule work as an aid to moral cognition and achieving an impartial perspective. He says of parents that they know how to give good gifts to their children because they provide them proper life-sustaining foods. By being available to their children in this way, parents affirm their own lives and those of their children.

Aristotle, like Jesus, used the example of parental love to help explain the nature of the link between self-love and love toward others. He finds this link in the relations characteristic of what he regards as genuine friendship: "Friendly relations with one's neighbors, and the marks by which friendships are defined, seems to have proceeded from a man's relations to himself."[25] Among these marks of friendship are: Friends wish one another to exist and live as mothers wish their children to exist and live; and friends grieve and rejoice with one another as mothers do with their children. So, as Aristotle goes on to indicate, the relationship we have to ourselves is of the very kind found in a mother's love for their children: We wish ourselves to live and exist; and we grieve and rejoice within ourselves. Furthermore, we wish good for ourselves. We do that as good individuals or as those who regard themselves as good. Existence and empathy are among the goods we wish for ourselves.

Both Aristotle and Jesus were using the example of parental love to specify what kind of relationships to ourselves are aids to moral cognition. If we have empathy toward ourselves of the kind parents exhibit toward their children, we will, like

---

[25] Aristotle, Nicomachean Ethics, book 9, Chap. 4.

parents, know what good we ought to bestow upon our friends and neighbors; we will have a basis for knowing how we ought to behave toward them. With empathy toward ourselves, we will above all wish them to exist and live as we wish ourselves to exist and live. In short, our moral cognition will be enlightened by doing to others what we would have them do to us. The Golden Rule is an aid to cognition if we love our neighbors as ourselves, and loving ourselves is understood as having our natural empathic emotions as the source of the relation we bear toward ourselves.

Loving our neighbors as ourselves is impartial in the sense that we extend to our neighbors the very same empathy we have for ourselves. We treat them as equals, as those to whom we owe the very same measure of justice and avoidance of injustice we expect from others. Our self-love, our empathic relations to ourselves, is impartial rather than self-partial.

Firth excluded self-love from what he regarded as impartiality because he thought of it as self-partiality, or at least as exclusively self-referential. Self-love would indeed be cognitively distorting, if the kind of love we have for ourselves were not to be at the same time the kind of love we have for others, and if we were to love only ourselves, or only some others. For our love to be impartial, it has to extend to everyone: Impartiality is open and inclusive, not closed and exclusive. We find this insistence on inclusion in the Good Samaritan story, Adam Smith's portrayal of an impartial spectator, H. R. Niebuhr's link of impartiality and universality, and in the omnipercipience attributed to an ideal observer by Firth, namely vividly imagining how actions affect *everyone*.

As we have characterized it, being impartial is being objective and being inclusive: These are distinguishable, yet related ways in which we attain an impartial cognitive perspective, a cognitively essential condition for being able to discern what we owe one another.

The reader will recall that we depicted achieving objectivity as an expression of impartiality, that is, as a matter of overcoming or obviating the distorting influence of *excessively* self-referential or self-partial particular interests and emotions. That the need for impartiality of this kind if justice is to be realized, is recognized in the practices of judges recusing themselves from cases in which they might not be able to overcome having their judgments influenced in their own favor by such particular interests or emotions present in the cases in question. To be objective, however, does require judges and others, not to be devoid of our natural empathic emotions, but that these emotions be inclusive. For judges to be impartial, requires that they be empathic both toward the alleged victim and the alleged perpetrator, attending to all the facts in such cases without regard to whether they benefit or harm either the plaintive or the defendant.

To be objective also necessities being inclusive in the sense of being informed by experience and perspectives other than our own. This helps us to overcome self-love that is too parochial and inadvertently self-partial. To trigger and expand the scope of our empathic emotions, for example, we increasingly ensure that African-Americans are represented on juries in trials involving African-Americans. We have similar policies with regard to gender.

At present, the role of empathy in our efforts to attain conditions designed to assure impartiality is not well understood. When Supreme Court Justice Sotomyer's appointment was being proposed, President Obama lauded her as a judge who has been, and would be, empathic. That prompted a chorus of strong objections insisting that judges are to be impartial, and that means being disinterested and dispassionate. No one objected to having an Hispanic and a woman on the Court because that would increase the mix that contributes to impartiality. However, that Sotomyer would be empathic was viewed as a red flag because it would or could mean that she would be biased in favor of women over men and Hispanics over others of a differing race or ethnicity. She was thoroughly grilled on the question of whether she would and could be impartial. No one, to our knowledge, surfaced with the idea that empathy is not only essential for making such decisions, but decisions are only impartial if equal empathy is extended to those who sue for justice and those who defend themselves as innocent of any alleged injustice. Without empathy, what is just cannot be discerned, and the urge for justice and the conscientious pursuit of it is not present. In a word, judges should be empathic, equally toward everyone, and inclusively of everyone. When they find themselves in situations that make it excessively and obviously difficult to be empathic equally and inclusively, they should recuse themselves and normally they do this without being asked or pressured.

This kind of sensitivity to impartiality, as discussed earlier, is present also in physicians. Indeed, research has shown that empathy is something patients seek from physicians, and for which they esteem them.[26] However, physicians would not be regarded as ideal, or ideally empathic, if they only showed empathy for some of their patients, certainly not by patients who were not treated empathically, nor by those who learned of such failures to manifest empathy toward all patients.

Impartiality, then, depends upon empathy and the Hippocratic ethic conforms to this very requirement of impartiality. Our natural empathic emotions incline us to be compassionate toward others and to do so in the ways we expect others to behave toward us. Also we do not wish to be excluded from being assisted when we have needs we cannot by ourselves meet. We cannot imagine that anyone else would normally wish or expect to be excluded in such circumstances either. We expect that physicians who have gained knowledge of and skill in providing medical care have this natural same moral basis for providing the care that is possible and necessary to everyone who needs it. As the Golden Rule makes explicit, they would not want or morally countenance being themselves denied medical care because they could not pay for it, and they can readily imagine that others in that situation share their wishes and moral outlook. Hence, there is this natural basis for the Hippocratic ethic and for the practice of Hippocrates himself in treating the poor.[27]

---

[26] N. M. Bendapudi, L.L. Berry, K.A. Frey, and W.L. Rayburn, "Patients' Perspectives on Ideal Physician Behavior," Mayo Clinic Proceedings 81.3 (March, 2006), 338–344.

[27] Oswei Tempkin, Hippocrates in a World of Pagan and Christians (Baltimore: The Johns Hopkins University Press, 1991), 220. On pages 216–223, Tempkin cites evidence for the practice of the Hippocratic ethic and its basis in love for humanity, compassion, reputation, and equity.

In summary, our account of the cognitive basis of moral decision-making began with the observation that the experience of the moral ought is that of relating a decision before us to our ideal self. Then we indicated that the notion of an ideal spectator or observer, or of an ideal moral judge, is one found in the philosophical literature and used to explain how we decide and verify or justify our moral actions or inactions. The most precise and developed description of such an ideal self is the one formulated by Roderick Firth. He describes an ideal moral judge as having the characteristics, the possession of which, specify by definition the conditions under which a given moral judgment is the correct one to make. These conditions consist of: Omniscience with respect to the non-moral facts, omnipercipience, impartiality, and the normal characteristics of a human being. He derived these from the cognitive processes that we are actually prone to set in motion in the course of moral decision-making.

However, since even as ideal selves we cannot attain the perfection necessary to guarantee that we are making correct moral judgments, we have to be content with making moral judgments that are the most plausible and verifiable as compared to their alternatives. That our moral judgments are less than certain to be true, holds for all of the judgments on which we base our knowledge of reality. Even knowledge in the most exact sciences is subject to changes for the sake of refinement and the correction of errors. To come as close to the truth and to what we count as convincing evidence for the claims we make, the cognitive basis for rationally justifying our moral decisions requires the following three processes: (1) A thorough, conscientious pursuit of non-moral facts related to choosing and implementing a policy; (2) a thorough and vividly imagined consideration of all who will be affected by any particular policy; and (3) an impartial perspective that is objective, inclusive, and equitable.

In this chapter, then, we have offered a cognitive basis for contending that our conception of what justice demands, as developed in the previous chapter, is rationally justifiable. In particular, we have defended the Hippocratic notion that everyone should receive medical care they need regardless of their ability to pay for it fully or at all. However, this explicit moral imperative is one that is commonly absent in the literature for what should be regarded as a just system of health care. The chapter now following presents an assessment of three prominent scholars who illustrate this widespread tendency to bypass the Hippocratic commitment to meet everyone's medical needs to see how they conceive of justice, and why their conceptions of justice prove to be inadequate.

Chapter 3
Advocating Basic Minimum Medical Care: A Case of Justice Denied

Abstract There is a widespread assumption that it is just to guarantee all individuals access to a basic minimum of healthcare. I contend in this chapter that this is based on a flawed concept and application of justice: it fails to use the generic concept of justice—namely of what we owe one another, that is, all moral demands, such as avoiding existing harmful practices and drawing upon the enormous amount of money generated by providing medical care to pay for that care. Omitting any references to these moral demands leads the proponents of minimal medical care to not even consider the possibility of achieving needed medical care for all. Thus they leave intact a system of healthcare that violates the impartiality justice demands—fair, equal treatment for everyone.

Currently, there are a number of notable scholars who consider universal access to medical care a demand of justice. To this proposition, our conception of justice, as we depicted it in Chap. 1, raises no objection. However, having *access* to those who provide medical care is not the same as receiving the care needed regardless of whether one is able to pay for it, and regardless of what the care costs. These scholars, and many others who advocate universal access to medical care, agree that what justice demands is "basic health care" guaranteed by the government.[1] In doing so, none of these advocates adhere to the demands of justice inherent in the traditional medical ethic that has guided medical care since at least the time of Hippocrates. The dominant major theories of justice neither conform to this Hippocratic medical ethic nor examine any reasons why anyone should.

Clearly the traditional medical ethic has a different conception of justice guiding medical practice. Its moral ideal does not condone denying needed care to anyone on the basis of insufficient means to pay for it. One would think the moral basis for this ethic would be examined, and any departures from it would have to be defended,

---

[1] In this chapter, we will analyze the works of three such notable scholars: Norman Daniels, Just Health: Meeting Health Needs Fairly (New York: Cambridge University Press, 2008); Martha C. Nussbaum, Frontiers of Justice: Disability, Nationality, Species Membership (Cambridge, M.A.: Harvard University Press, 2006); Amartya Sen, The Idea of Justice (Cambridge, M.A.: Harvard University Press, 2009).

by those who analyze and propose what counts as justice in what medical care should be provided universally. The departure from the traditional medical ethic is not confined to treatises, however scholarly or journalistic. It is enshrined in policy as well. The American Medicare program, however generous one may deem its benefits to be, is a prime example. The official government handbook is very explicit about this:

> Medicare doesn't cover everything. If you need certain services that Medicare doesn't cover, you will have to pay for them yourself unless you have insurance to cover the costs... Even if Medicare covers a service or item, you generally have to pay deductibles, coinsurance, and copayments.[2]

State governments, with financial assistance from the federal government, administer Medicaid, a program that allows impoverished persons to access and receive some medical services. In this instance as well, the services provided for are limited so that in states like Oregon, some services are explicitly excluded from reimbursements. This means that, as in the case of Medicare, those with sufficient means may, through the purchase of insurance and/or out-of-pocket payments, have their medical needs met fully or at least to a much greater extent than those of modest means or those mired in poverty. The health care reform bill passed in 2010 will expand access and coverage for a greater number of individuals and families.[3] It is not, however, governed by the ideal of the traditional medical ethic: One's level of income remains a significant factor in determining the level and extent of medical services that one will be able to receive.

Unquestionably, then, how we think about justice is not a matter of indifference. Different views of justice, however explicitly or implicitly, result in different social policies. It is evident that the traditional medical ethic reflects an understanding of justice that receives no explicit attention in current theories of justice generally, nor in legislative action to assist individuals and families to access and obtain medical care. Our notion of justice, undergirding the traditional ethic, will not be accepted without a stout and convincing defense. We will mount one in this chapter.

In order to make the case for retaining and expanding the traditional medical ethic, it will be necessary to receive, refine, and render more comprehensive our account of justice. At the same time, it will be necessary to consider why the currently influential theories of justice espoused by Norman Daniels, Martha Nussbaum, and Amartya Sen do not provide support for the traditional medical ethic, and why they fail even to take it into account as a view of what justice demands when it comes to the provision of medical care.[4] Furthermore, for various reasons we will offer, these theories do not sufficiently justify their views of justice and its applicability to formulating and shaping social policy, particularly health care policy.

---

[2] Medicare and You (U.S. Department of Health and Human Services, 2011), 46.
[3] The Patient Protection on Affordable Care Act (On Hundred Eleventh Congress of the United States of America: H.R. 3590, January 5, 2010).
[4] Daniels, Just Health; Nussbaum, Frontiers of Justice; and Sen, The Idea of Justice.

3.1 The Moral Imperative of Justice

To address the question of what constitutes justice in the provision of medical care in Chap. 1, we examined the concept of justice in its universal, generic sense. Justice, in its general sense, refers to what we owe one another. We found that justice in this sense reveals the moral basis for the traditional medical ethic. That way of thinking about the moral imperatives guiding the provision of medical services puts us in partial, but highly significant opposition, to the views currently regnant among those who also advocate universal access to medical care. These advocates of universal access to medical care do not base their support for it on a generic sense of justice; for them, and that includes Daniels, Nussbaum, and Sen, justice is equated with distributive justice and in a way more narrowly conceived.[5] In what follows, we shall see that viewing justice as they do bypasses the moral basis for the Hippocratic ethic.

As we noted in Chap. 1, justice, what we owe one another, is expressed in the moral responsibilities we as human beings naturally exhibit in making it possible to bring into being and sustain individuals and communities. Nurture is one such moral requisite of individual and communal life. It refers to a whole range of responsibilities experienced as morally obligatory and aimed at sustaining and protecting human life. These responsibilities include efforts to prevent illness and suffering, and provide care for those who are ill, disabled or otherwise in distress, and rescue those at risk of losing their lives were no one to come to their aid. We have natural proclivities to behave in these ways.[6]

We also are naturally inhibited when it comes to harming and killing others. These felt moral responsibilities are also requisite for the existence and persistence of individual and communal life. Other such moral requisites include our inhibitions about lying and stealing which, if absent or uncontrolled, would subvert and destroy individual and communal life.[7]

Individuals cannot on their own assure that they and others will be able to nurture and receive nurture, and prevent harms that result from lying, stealing and the threats to well-being and life from illnesses and disasters, whether accidental, natural, or manmade. These necessities spawn the various institutions and professions that have the ability to meet these responsibilities or assist individuals, families and

---

[5] One notable exception to the practice of finding the moral basis for universally accessible healthcare in a concept of distributive justice is the appeal to beneficence made by Allen Buchanan, "Health-Care Delivery and Resource Allocation," in Robert M. Veatch, ed., Medical Ethics (Subury, M.A.: Jones and Bartlett Publishers, 1997), 321–361.

[6] As noted in Chap. 1, the neurosciences provide evidence for such natural proclivities. For this, see Martin L. Hoffman, Empathy and Moral Development: Implications for Caring and Justice (New York: Cambridge University Press, 2000); and Allan N. Schore, Affect Regulation and the Origin of the Self: The Neurobiology of Emotional Development (Hillsdale: N.J.: Lawrence Erlbaum Associates, 1994), 348–354.

[7] Hoffman, Empathy and Moral Development, refers to work with juvenile delinquents that uses techniques to evoke empathy and thus awaken inhibitions against aggressive and violent behavior.

other groups to do so. Thus we have governments, laws, lawyers, police forces, armed forces, fire departments regulations and agencies to assure safe foods, medicines, air, and water, and professions dedicated to care, healing, and the prevention of various threats to well-being, professions such as physicians, nurses, dentists, chiropractors, nutritionists, pharmacists, and counselors, trained as psychiatrists, psychologists, and religious clerics. Be they ever so various, cultures are not without those who qualify and serve as healers. When we speak of medical care in this treatise on justice, our focus is on the services rendered by physicians, while recognizing the vital roles of nurses, medical technicians, pharmacists, and nutritionists among others, to meet the goals of the medical services being rendered by physicians.

The moral imperatives of justice, particularly nurture as care and rescue and the prevention of harm, account for and make sense of the practices and ethical impulses of physicians. Threats to health, life, and our abilities to meet our goals are distresses that invoke the empathetic emotions and motivate some to acquire the skills and knowledge to aid us when faced with such distresses. What constitutes health and threats to it are by and large defined both by those seeking assistance and these providing it. Once individuals have sought out a physician, what constitutes health for such individuals becomes mostly the province of the physician. What remedies are appropriate and necessary, and whether any are necessary at all are determinations that largely flow from the physician's notion of health and the threats to it. The moral imperatives of justice we have identified, and that shape the Hippocratic ethic, also call for offering only the services deemed to be appropriate and necessary and doing so even for those who can pay little or nothing for these services. To exclude anyone from the care medically indicated would clearly go contrary to the physician's empathic emotions and would tend to induce feelings of guilt and/or shame.[8]

When, then, we ask what we owe one another in the matter of medical care, it is the moral imperatives of justice in this most generic sense that govern how the distribution of medical care is structured and who distributes it. The moral impulses of justice move people to seek medical care and move individuals to acquire the knowledge and skills to undertake that care. These same moral impulses have also motivated religious institutions, foundations, and governments to help finance medical infrastructure, education, and research essential to improving and professionalizing the practice of medicine.

Among the moral impulses of justice directly tied to how goods and services are distributed is that of fairness. It does not take long for children to detect unfairness. They are quick to protest receiving a smaller slice of a freshly baked pie than the portions being given to others, particularly their peers. As the philosopher William Frankena has so well stated it:

> The paradigm case of injustice, if there is one, is that in which there are two similar individuals in similar circumstances and one of them is treated better or worse than the other. In

---

[8] That shame and the anticipation that certain behaviors would be shameful is an essential ingredient in moral development, is documented by Schore, Affect Regulation and the Origin of the Self, 349–350.

this case the cry of injustice rightly goes up against the responsible agent or group; and unless that agent or group can establish that there is some relevant dissimilarity after all between the individuals concerned and their circumstances, he or they will be guilty as charged.[9]

In short, justice is the similar and injustice the dissimilar treatment of similars. Frankena regards this as a necessary condition of distributive justice, or as we prefer to describe it, as one of the moral imperatives of justice. Clearly, empathic physicians would wish to avoid the rightly anticipated distress and harm of turning away individuals who cannot pay for the care these physicians have determined they need. That means that medically similar cases are treated similarly regardless of whether they involve people who are rich or poor. That is a moral imperative embedded in the traditional medical ethic, namely being fair.

The late philosopher John Rawls championed justice as fairness and his theory has greatly affected the thinking of leading proponents of universal access to medical care.[10] However, this has not led to support for the traditional medical ethic as will soon become evident for reasons we will now explore and assess.

3.1.1 Justice as Fairness Espoused by Norman Daniels

Among those who have embraced a Rawlsian theory of justice, Norman Daniels stands out as one who has formulated what is arguably the most plausible and influential account of how the theory applies to the realm of providing medical care. The subtitle of his most recent book, namely "Meeting Health Needs Fairly," neatly summarizes its orientation and goal.[11]

It is important to recognize from the outset, that when Rawls and Daniels speak of justice, they are referring to distributive justice. They do not use the term justice in its generic sense to describe the whole range of our moral responsibilities, what we owe one another as human beings, fairness to be sure, but much more than that. For them, justice refers exclusively to fairness and in particular to a fair distribution of social goods.

Rawls identified the following as primary social goods: basic liberties, powers, prerogative of office, income and wealth, the social basis of self-respect, and opportunities.[12] Rawls articulates two principles that comprise the moral demands of justice and which are to guide decisions aimed at realizing a just distribution of these primary social goods. The first principle is that: "Each person is to have an equal right to the most extensive total system of liberties compatible with a similar system

---

[9] William K. Frankena, Ethics (Englewood Cliffs, N.J.: Prentice-Hall, Inc., 1963), 39.

[10] John Rawls, A Theory of Justice (Cambridge, M.A.: Harvard University Press, 1971). For his final work, edited by Erin Kelly, see Justice as Fairness: A Restatement (Cambridge, M.A.: Harvard University Press, 2001).

[11] Daniels, Just Health.

[12] Rawls, A Theory of Justice, 92; Rawls, Justice as Fairness, 174.

of liberty for all."[13] The second principle, the "difference principle" is that: "Social and economic inequalities are to be arranged so that they are both (a) to the greatest benefit of the least advantaged and (b) attached to offices and positions open to all under conditions of fair equality of opportunity."[14] The general conception of justice is this difference principle "applied to all primary goods including liberty and opportunity."[15] However, the two principles in serial order are "the form that the general conception finally assumes as social conditions improve."[16]

It is this formulation of the moral imperatives of justice that Daniels draws upon to answer what he refers to as the "'Fundamental Question' of justice, for health."[17] The question is the following one: "As a matter of justice, what do we owe each other to promote and protect health in a population and to assist people when they are ill or disabled?"[18] The obligations being asked about are for Daniels social obligations. To meet these obligations will require governmental actions and policies. That is the case for the primary social goods Rawls has identified as such. However, Rawls did not explicitly include health among the primary social goods. Daniels, then, sees the need to argue for the special moral importance of health so that what should be done to provide and protect it is indeed a matter of justice and, as such, a social obligation involving governmental actions and policies.

Daniels defines health as "normal functioning for our species."[19] As such, health contributes to protecting opportunity since normal functioning is what allows people to maintain a normal range of opportunities. That, for Daniels, establishes the special moral importance of health because protecting opportunity is a requirement of justice. Daniels summarizes his central argument for the moral imperative to meet health needs as follows:

> (1^1) Since meeting health needs promotes health (or normal functioning), and since health helps to protect opportunity, then meeting health needs protects opportunity. (2^1) Since Rawls's justice as fairness requires protecting opportunity, as do other important approaches to distributive justice, then several recent accounts of justice give special importance to meeting health needs.[20]

Daniels, of course, includes health care (medical care) as one of the means of meeting health needs. It follows, then, that the provision of medical care is among the moral demands of distributive justice. Medical care is not a primary good; opportunity is. However, health care needs are connected to the institutions that provide for fair equality of opportunity and that makes meeting them a requirement of distributive justice.[21]

---

[13] Rawls, A Theory of Justice, 250.
[14] Ibid., 83.
[15] Ibid.
[16] Ibid.
[17] Daniels, Just Health, 11.
[18] Ibid.
[19] Ibid., 37.
[20] Ibid., 30.
[21] Ibid., 56–58.

3.1 The Moral Imperative of Justice

Daniels acknowledges that people, if asked about the importance of medical care, will give many reasons, including that it saves lives or reduces suffering. Nevertheless, his claim is that, his explanation as to why medical care is important reflects people's real interests. Furthermore, he asserts that his explanation "coheres well with the justification for thinking that meeting health needs is morally important, namely that we have an obligation to protect opportunity and that protecting health protects opportunity."[22]

On the face of it, linking health to a normal range of functioning, and to having a normal range of opportunities to choose from, would appear to be unproblematic. Indeed, there is no reason to dispute that there is a connection between being healthy and being able to take advantage of one's opportunities. However, linking the protection of health with the protection of opportunity from the standpoint of a Rawlsian theory of justice is quite another matter, for opportunity is one good among others that the government is obligated to distribute and guarantee with its financial support. Daniels is explicit about this: "Despite the importance of opportunity, we must limit our spending on health care…opportunity is only one of many important goods."[23]

Seen from the perspective of a generic conception of justice, Daniels is definitely understating the moral significance of medical care. Working without a generic concept of justice, Daniels does not attend to or identify the most salient moral reasons individuals have for seeking and expecting to receive the care they need, and physicians are motivated to provide just that. When people say that medical care is important because it saves lives and relieves suffering, they are giving moral reasons for its considerable importance. Saving lives and relieving suffering are moral responsibilities, requirements of justice perceived and motived by our natural empathic emotions: This is what justice as nurture urges upon us as human beings. At the same time, nurture is a moral requisite of individual and communal life. Seeking and providing care, and saving people from dying and the risk of dying, are requisites of having the opportunity to meet our moral responsibilities toward ourselves and others, as members of various communities, such as our families and nations, and ultimately of the whole human community.[24] In short, caring for our health is a moral responsibility for everyone, and a special calling for physicians and other health care practitioners. It is not wrong to associate health care with the protection of those functions essential to pursuing opportunities. However, what makes doing this a matter of special moral importance is that maintaining our health is a moral responsibility in itself and necessary if we are to meet our other moral responsibilities. Indeed, our very lives are at stake. That is surely of the utmost importance and the obvious requisite of being able to pursue any opportunity at all.

Defending universal access to health care without recourse to a concept of generic justice, then, leads Daniels to ignore or overlook the moral demands of jus-

---

[22] Ibid., 30, Footnote 1.

[23] Ibid., 63.

[24] For a definition of community and universal community, see Arthur J. Dyck, Rethinking Rights and Responsibilities: The Moral Bonds of Community (Washington, D.C.: Georgetown University Press, 2005), 95 and 199–200.

tice underlying the traditional medical ethic, allowing him to adopt a view of distributive justice that does not meet the standard of fairness found in that ethic, a standard he does not even discuss. Having understated the special moral importance of medical care, he compares the obligation to provide medical care with the obligation to provide education. The needs for both are connected to the institutions that provide for fair equality of opportunity and it is opportunity that is the primary social good, not medical care or education.[25] For these, as for opportunity as one primary good among others, the obligation of distributive justice is to set and attain a decent social minimum. Hence, as Daniels says, "A just system must provide for more than health care. Justice stops us from falling into a bottomless pit of health needs."[26] Medical care as a matter of justice must be limited: What then does justice demand? Daniels addresses that question as follows:

> Personal medical services considered essential to promoting fair opportunity for all must be accessible to all. This will generally mean universal coverage through public or private insurance for an array of "decent" or "adequate" services in order to protect fair equality of opportunity. There should be no obstacles—financial, racial, geographical, and so on—to access to the basic tier of the system. Determining what is in that basic tier must be clarified in light of arguments about how to protect fair equality of opportunity under reasonable resource constraints; these arguments require a fair process (accountability for reasonableness) for appropriate democratic deliberation...[27]

Assuming as he does that it is reasonable to deny treatment to someone for financial reasons, Daniels discusses a hypothetical case in which health authorities or health plans make conflicting decisions, that is, different coverage decisions for specific patients under identical conditions. Both Jack and Jill and their physicians agree on the need for a specific treatment regimen. This treatment has not been conclusively proved to be of benefit for the condition in question. However, the standard treatments cannot prevent people who have this condition from dying because of it. Daniels supposes further that: "After careful deliberation, weighing such relevant reasons as the importance of conserving resources against compassionate use of unproven therapies, one health authority (or health plan) approves the regimen for Jack and the other denies it to Jill."[28] On the basis of this case, Daniels then poses the question as to whether we should agree, if Jill complains of unfair treatment she has indeed been treated unfairly.

First, Daniels considers that Jill might claim that justice has not been served because the formal principle that like cases be treated alike has been violated. Since Jack and Jill have the same malady and prognosis, either both or neither should receive the only possible intervention. If there are reasons to treat Jack, the same reasons apply to her.

Daniels does not regard this formal principle as a compelling reason for concluding that Jill was unfairly treated. He presupposes that the procedures followed in

---

[25] Daniels, Just Health, 57.
[26] Ibid., 63.
[27] Ibid, 143–144.
[28] Ibid., 135.

3.1 The Moral Imperative of Justice

managing the cases of Jack and Jill were carried out in a reasonable manner. The difference in the two cases rested on a difference in the weights given to certain values such as that of "urgency, stewardship, and shared decisions making, with patients."[29] Daniels supposes further that the weighting itself is a matter of reasonable disagreement and there is no argument that acceptably shows "that one weighting of these values is clearly more morally justifiable than the other."[30] There is, then, uncertainty about what amounts to a just outcome and we are compelled to adopt a procedural approach to fair outcomes. Jill, therefore, can be told that Jack's plan reasoned differently than your plan and "both ways of reasoning are relevant and arguably fair."[31]

Daniels takes up the possibility that Jill might sue her health plan. Although the courts might make experiments with different fair procedures unworkable, he inclines to the view that the courts might decide a case like Jill's on the basis of whether or not fair procedures were followed. In short, Jill could lose in court.

In the end, Daniels sees the acceptability of differential treatment dependent upon whether a political rationale for uniformity can be developed that would prove persuasive. He finds that to be a difficult proposition. He concludes that: "For the problem we are facing, then, it remains unclear how unacceptable it would be for Jack to get a last chance when Jill does not."[32]

The way in which Daniels approaches the hypothetical cases of medical decisions made for Jack and Jill illustrates very well critical differences between his conception of justice and the one we are defending. In this very significant case as he has constructed and discussed it, the inadequacies of his view of justice become very apparent. This case of conflicting decisions should not be reduced to a procedural question. Certainly decision-making procedures should be reasonably, accountably, and fairly conducted. However, what justice demands should be done for both Jack and Jill is the question that should and can be addressed reasonably, accountably, and fairly.

First of all, Daniels does not explicitly invoke a duty to rescue even though both Jack and Jill have only this one last chance at life, namely the intervention being offered to Jack but not to Jill. He does mention urgency as one value to be considered. However, he allows for a fair process that would weigh more heavily the conservation of scarce resources and not leaving the decision to Jill. This is not the prevailing way to regard the weightiness of trying to save a life. When there is an explosion in a mine or a destructive earthquake, the first question about resources is not how to conserve them, but what is needed and how to procure them as speedily as possible so that anyone trapped by one of these disasters may be rescued. Efforts to rescue continue as long as there is even a remote chance that these may save a life. What weighs against a rescue effort is the risk to those of the rescuers or those trapped. When a life is at stake, it is not the financial cost that is being counted, but

---

[29] Ibid., 136.
[30] Ibid.
[31] Ibid.
[32] Ibid., 137.

the incomparable worth of life, in the decisions being taken to save a life. Furthermore, the assumption is that people wish to live. When the decision is made to undertake a recue it is also a decision to keep up the effort until all, not some, are accounted for, and until the effort is deemed to be utterly futile, or for some reason has become too dangerous. Rescuers might differ as to when a rescue is not feasible or too dangerous. They would not differ about making the same attempt to save Jill as they do to save Jack when they are found to be similarly situated, and the attempt, however desperate, is given a remote chance of success and the decision is made to undertake it. When it comes to saving lives, like cases are to be treated alike. Our empathic emotions, the source of our capacity to be moral, to be just in the generic sense, propels us to act in this way.

Although Daniels does not explicitly invoke a duty to rescue, he does recognize that the impulse to grant someone like Jack a last chance at life is a compassionate response to Jack's plight. He describes the deliberations as to whether to treat Jack and Jill as "weighing such relevant reasons as the importance of conserving resources against compassionate use of unproven therapies."[33] However, he does not equate a compassionate act with a just act. For him, it can be just to deny care for the sake of conserving resources in this instance. From our standpoint, if this offer of therapy to Jack and Jill is compassionate, it is just, and because it is just, resources if scarce are to be sought, not withheld. If the intervention contemplated for Jack and Jill were to be judged futile by their *physicians*, it would not be compassionate to put them through the rigors of the chemotherapy contemplated in this case; the treatments would be unjust, both as a misuse of resources and as lacking in empathy, in compassion.

Daniels might well reply, that even if we were to regard a compassionate act as a just act, conserving resources also meets the demands of justice, and that demand must be weighed against the other demands of justice: Resources expended in the medical care sector should not expand to such an extent that it becomes impossible to sustain a just distribution of primary goods. The reality of finite resources must be confronted. A disaster, if extensive enough, may exceed our resources adequately to comply with the duty to rescue. The appeal to this imperative simply does not address the problem of justly distributing goods with finite resources.

If Daniels were to claim that the duty to rescue does not, or cannot, guide the distribution of scarce, even insufficient medical resources, he would be mistaken. Combined with treating like cases alike, the moral imperative to save lives that are at risk, is always relevant to medical decision-making. To begin with, physicians, in the course of rendering their services, are always more or less scarce resources. Those who can wait for their services are treated as sufficiently able to be assigned appointments on a first-come, first-served basis. When a physician encounters an emergency case during a given schedule of appointments, one or more appointments will be cancelled and/or postponed as deemed necessary and non-threatening to those who must delay seeing their physicians. The moral imperative to rescue takes precedence and the scarce resources are made available as needed to carry out

---

[33] Ibid., 135.

3.1 The Moral Imperative of Justice

the rescue. There are also scarce resources such as kidneys for saving and/or extending the lives of patients with failing kidneys. In situations like this, the duty to rescue motivates people in both the private and public sectors of society to make efforts to increase the number of kidney donors, and of those who will permit others to be donors or have their kidneys donated. At the same time, less costly means of keeping people alive and functioning with dialysis have been very creatively invented and utilized.

Without this kind of link between the duty to rescue and the moral demands of justice, including distributive justice, Daniels concentrates on the alleged necessity to limit therapies even those that might save the life of someone like Jill. What is missing in his approach to the limits of resources is an analysis of how resources could be expanded without adversely affecting the distribution of primary goods and without denying appropriate medical care. That is the way medical needs are to be met when the distribution of medical care is guided by the traditional medical ethic, and when the moral imperatives of justice in its generic sense, embedded in that traditional medical ethic, is what undergirds and motivates it practitioners. Furthermore, without attending to the empathic emotions that move us to share and work for the resources needed for universal care, and without seeing empathy as the source of justice, Daniels is left without a compelling reason to consider how the traditional medical ethic could be made to work in our present circumstances. Since, given our view of justice, the Jills of this world are owed the same chance at life that it is deemed reasonable to offer to the Jacks of this world, we do have the strong moral imperative to examine the resources available in the medical care system of the United States, one that generates and spends more money than any other nation in the world.[34]

Daniels is certainly aware that people generally are strongly opposed to denying care on the basis of how much it costs, especially denying life-saving care because it is regarded as too costly. As he notes, in cases of denying care when "slightly greater net benefits are possible, especially the decreased risk of death, but only at a much greater cost, there is a strong risk of being criticized for putting so direct a price on the value of life."[35] He refers to two examples of refusing to cover certain expensive drugs in Britain and Canada.[36] What he observes about these examples, however, is that they foster disagreement, and that they illustrate that there are health authorities using cost-effectiveness to make decisions about coverage. Fear of criticism leads private plans in the United States and the Center for Medicare and Medicaid Services to refrain from publically embracing cost-effectiveness as a criterion for their decisions.[37] Despite this acknowledgement of resistance to cost-

---

[34] See, for example, Peter A. Muennig and Sherry A. Glied, "What Changes in Survival Rates Tell Us about U.S. Health Care," Health Affairs 29:11 (November, 2010), 2105.

[35] Daniels, Just Health, 127.

[36] Ibid.

[37] Ibid., 128. However, there is some support for cost-effectiveness in the U.S. Congress. Cost-effectiveness appears as a criterion for care-giving in the Patient Protection and Affordable Care Act.

effectiveness, Daniels maintains that cost-effectiveness should count among the relevant reasons to deny marginally effective but quite costly treatment even if that treatment is the only one available.[38] The consideration of costs becomes problematic, however, if it does not occur in decision-making guided by all of the moral imperatives of justice. Daniels is setting aside, or at least not exploring the various fears and objections being raised by the use of costliness to deny medical care. Furthermore, his understanding of justice does not lead him to indicate how and why medical interventions are limited justly for the benefit of those who seek medical services. Our concept of justice addresses both of these vital matters.

The use of costliness as a criterion for denying care threatens to undercut the deepest reasons people have for availing themselves of medical assistance: They do not wish to die; they do not wish to suffer unnecessarily. These are the moral imperatives of nurture and rescue, moral imperatives of justice. From the perspective of the medical practitioners, denying nurture and rescue when the need for them is medically indicated and therapeutically possible to meet, runs counter to their empathic emotions, the source both of their impulses to be just, and of their understanding of what justice demands of them. The urges to offer relief to those who are painfully suffering, or to offer aid to someone at risk of dying, are extremely compelling, so much so that gross deviations from such urges are virtually beyond comprehension. That Nazi physicians could regard some human beings who required care as not worth the cost of caring for and sustaining them, not only shocks us, but also has inspired numerous studies aimed at trying to explain and ward off such behavior in the future. Among these scholarly analyses is that of Robert Jay Lifton who concluded his book with these admonitions for physicians: (1) Maintain a balance between ethical commitments and technical skills; (2) Be committed always to Hippocratic principles of healing; (3) Be critically conscious of any large projects demanding one's allegiance; (4) Cultivate an "equally pervasive empathy, fellow-feeling, toward all other human beings."[39]

Daniels, of course, would repudiate the atrocities perpetuated by Nazi physicians. However, in the name of justice he should take more seriously how threatening it is to invoke cost to deny treatment to anyone, most particularly to deny life-saving treatment. The worth of a human life is beyond calculation. Similarly, he should not compromise the commitment to the equal worth of lives: Justice demands strict equality of opportunity for life-saving interventions in situations like that of Jack and Jill. Our concept of justice calls for what Lifton affirms, namely "equally pervasive empathy, fellow-feeling, for all human beings."[40]

Given his particular theory of justice, there is another moral imperative of justice in its generic sense that Daniels does not take into consideration. To avoid harming one another is a stringent demand of justice in our relationships. It is a time-honored guide for practicing medicine in a morally responsible way. First of all do no harm!

---

[38] Daniels, Just Health, 128.

[39] Robert Jay Lifton, The Nazi Doctors: Medical Killing and the Psychology of Genocide (New York: Basic Books, Inc., 1986), 500.

[40] Ibid.

3.1 The Moral Imperative of Justice

This demand of justice clearly obligates physicians to limit care. Edmund Pellegrino describes this morally obligatory kind of gatekeeping, a role traditionally binding for medical practitioners, as follows:

> They must use their knowledge to practice competent, scientifically rational medicine. Their guidelines should be diagnostic elegance (i.e. using the right degree of economy of means in diagnosis) and therapeutic parsimony (i.e. providing just those treatments that are demonstrably beneficial and effective). In this way, the physician automatically fulfills economic and moral obligations. He or she simultaneously avoid unnecessary risk to the patient from dubious treatment and conserves the patient's financial resources, and society's as well.[41]

This kind of gatekeeping is totally for the patient's benefit. As Pellegrino rightly notes, however, there are gatekeeping activities that violate the best interest of patients. This occurs when they "deny needed services or induce patients to demand, or the physician to provide, unneeded services."[42] These moral failures to treat patients justly arise in a context in which what is necessary and unnecessary care is not determined by physicians but by formulas formulated and applied by hospitals and/or other health authorities, whether private or public, and by "Medical criteria subject to modification or veto by economic considerations."[43]

Our concept of justice supports Pellegrino's delineation of what constitutes morally justified and morally unjustifiable denial of medical care. Patients are owed care deemed necessary on the basis of medical, not economically modified determinations of care, albeit in the least costly way of securing the best interests of patients. Practicing medicine justly does limit care by avoiding excessively risky, insufficiently beneficial and/or unnecessary and unnecessarily costly interventions, and doing so for the well-being of patients, not for the sake of reducing costs.

When Daniels declares that justice keeps medical needs from becoming a bottomless pit, he does not refer to the morally justifiable limits to care we have been describing. Rather, his theory of justice calls for limiting medical care on the basis of economic considerations as a matter of policy, policies set and administered by health care authorities and insurers, whether public or private. At the same time, financial support for medical care, as a matter of policy, is limited to a so called "decent minimum" that does not include all the care physicians may regard as necessary, on generally accepted medical standards, competently and conscientiously judged on a case by case basis.

Unless Daniels can demonstrate the strict necessity for adopting the policies he advocates, he is advocating practices that violate the demands of justice as we understand them and find to be embedded in the traditional medical ethic. It is not enough to assert the finitude of medical resources. Our use of medical resources is also limited, and justly so, by the practice of scientifically rational, efficient, and

---

[41] Edmund Pellegrino, "Rationing Health Care: The Ethics of Medical Gatekeeping," in Health Care Ethics: Critical Issues for the 21st Century, John F. Monagle and David C. Thomasma, eds., (Gaithersburg, Maryland: Aspen Publishers, Inc., 1998), 415.

[42] Ibid., 416.

[43] Ibid.

compassionate care. Justice can be even better served by increasing the incentive to be just in all of these ways. Furthermore, Daniels has not provided any evidence that current medical resources are, or will be, insufficient for practicing medicine in accord with the traditional medical ethic.

3.1.2 The Capabilities Approach to Justice in Martha Nussbaum and Amartya Sen

Like Daniels, Nussbaum and Sen advocate universal access to medical care that guarantees an adequate or basic minimum of care. What counts as adequate or basic is left to political, publically accountable processes to determine. As we indicated at the outset of this chapter, none of these theories of justice is in accord with the Hippocratic medical ethic, nor do they offer any reasons why anyone should be. Similarly, none of these theories examine justice in its generic, comprehensive sense, that is, as a concept that identifies and describes the various morally significant relations in which we stand as human beings, relations that constitute moral requisites of individual and communal life.

Nussbaum comes close, at times, to thinking of justice as moral responsibilities that exist in and make possible, individual and communal life. Drawing upon the thinking of Hugo Grotius, she declares that: "Wherever human beings are alive, there are already circumstances of justice between them, just because they are human and sociable."[44] Again, at the point of introducing her list of capabilities, she explains that her version of the capabilities approach begins "with a conception of the dignity of the human being, and of a life that is worthy of that dignity—a life that has available in it 'truly human functioning'..."[45] To have such a life, entails for Nussbaum, that citizens have certain capabilities that provide for them a range of opportunities for activity. Justice demands these as entitlements. She then describes her list of capabilities as "central requirements of a life of dignity."[46] Among these capabilities are life, bodily health, bodily integrity, and affiliation, social and political.[47] These certainly constitute what one can call requisites of sustaining individual and communal life. However, her discussion of capabilities is such that they are not being characterized or justified as *moral* requisites of individual and communal life, nor as arising from them.

Nussbaum presents her theory of justice as an alternative to utilitarian and contractarian theories, including that of Rawls. To begin with, she believes "that the capabilities approach supplies sounder guidance for law and public policy."[48] Her

---

[44] Nussbaum, Frontiers of Justice, 38.
[45] Ibid., 74.
[46] Ibid., 75.
[47] Ibid., 76–77.
[48] Ibid., 70.

theory is designed to offer a philosophical basis "for an account of core human entitlements that should be respected and implemented by the governments of all nations as a bare minimum of what respect for human dignity requires."[49] Human capabilities refer to "what people are actually able to do and to be."[50] The goal is to guarantee that each individual citizen's capabilities are supported sufficiently to make a life of dignity, of truly human functioning, possible. Thus her list of capabilities comprises a list of entitlements. These relevant entitlements "are prepolitical, not merely artifacts of laws and institutions" such that "a nation that has not recognized these entitlements is to that extent unjust."[51] In short: "The sphere of justice is the sphere of basic entitlements."[52] The capabilities approach specifies appropriate political goals. Indeed, capabilities are political principles, selected for political purposes only, as guides to laws and policies. The capabilities approach does not purport to be "a complete theory of justice" but rather provide a list of capabilities that "give us the basis for determining a decent social minimum in a variety of areas."[53] Since, as noted in the previous paragraph, life, bodily health, and bodily integrity are on her list of capabilities that should guide laws and public policies, a decent minimum of medical care for each individual citizen is what, in her judgment, governments should not fail to provide.

Unlike Daniels, Nussbaum does not raise the issue as to whether a given government has the resources to assure that all of its citizens are accorded a decent minimum of health care. Capabilities are "social goals" that governments should strive to actualize to the level that permits each citizen to live a life of dignity, that is, to possess the capabilities necessary for truly human functioning.[54] Furthermore, for adult citizens, *"capabilities, not functioning, is the appropriate political goal."*[55] The reason for this is that practical reason, a very important capability on Nussbaum's list, is what makes all the other capabilities on the list human, making it possible to choose the functions in accord with some conception of the good that allows for a life of dignity.[56] Of course, Nussbaum recognizes that no government can guarantee that all of its citizens will be healthy or free of illness. Therefore, she describes the goal with respect to health as that of achieving a social basis for it. The guarantee of a decent minimum of medical care for all its citizens is one of the necessary steps toward realizing the social basis for health. Having medical care available to them renders individuals more capable of attaining and maintaining health, and dealing with illnesses. Nussbaum also describes this capability as an opportunity "to lead a

---

[49] Ibid.
[50] Ibid.
[51] Ibid., 285.
[52] Ibid., 337.
[53] Martha C. Nussbaum, Women and Human Development: The Capabilities Approach (New York: Cambridge University Press, 2000), 75.
[54] Ibid., 86.
[55] Ibid., 87.
[56] Ibid.

healthy lifestyle," one that leaves people free to make choices in matters that pertain to their health, even to make unhealthy choices.[57]

Daniels would see Nussbaum's reference to a capability as an opportunity, as a recognition, however inadvertent, that his theory of justice, and those of Nussbaum and Sen, differ in terminology only and not in substance when it comes to what the targets of justice should be. Nussbaum and Sen explicitly reject primary goods as the goals of justice in favor of capabilities. Opportunity is one of those primary goods. As Daniels effectively argues, the social minimum that Nussbaum and Sen seek for capabilities, leads them to propose adequate medical care for every citizen, a proposal mirroring his own, and adequacy remains undefined for them as it does for him.[58] Similarly, when it comes to what medical care does for people, the capabilities approach appears to be aiming for something like the normal range of opportunities as specified in Daniels. However, unlike Daniels, Nussbaum and Sen do not treat resources such as income and wealth as primary goods; rather they regard them as means to achieve the social minimum of the various capabilities and not as competing with them as demands of justice as such. Nussbaum explicitly rejects the balancing of competing ends found in the Rawlsian approach. Entitlements cannot substitute for one another. The whole set is required of justice: "The approach positively forbids...trade-offs and balancing when we are dealing with the threshold level of each of these requirements."[59] An earlier version of this stand against trade-offs, mentions how this limits the use of cost-benefit analyses, and any analysis that would push any entitlement below its threshold should be viewed as a tragedy, one that cannot be compensated by other gains such as economic growth.[60] The goal to bring that capability up to its required level remains in place as a strict requirement of justice.

With respect to medical care, it is true that Nussbaum does not define adequacy, the threshold of the social minimum for it. However, the provision of medical care and education, primary and secondary, is so important that adequacy requires "something close to equality, or at least a very high minimum."[61] Of course, that still leaves this threshold to be determined politically. However, Nussbaum warns governments to take seriously the urgency of assuring adequate medical care or face dire consequences: The desire for health is such a strong, permanent feature of our makeup as humans, regimes that fail to respond adequately to this desire and need are, or will be unstable.[62] From a moral perspective, she characterizes a failure to promote adequate medical care "as causing a kind of premature death, the death of a form of flourishing that has been judged to be worthy of respect and wonder."[63]

---

[57] Nussbaum, Frontiers of Justice, 79–80.
[58] Daniels, Just Health, 64–71.
[59] Nussbaum, Frontiers of Justice, 85.
[60] Nussbaum, Women and Human Development, 81.
[61] Nussbaum, Frontiers of Justice, 294.
[62] Nussbaum, Women and Human Development, 155.
[63] Nussbaum, Frontiers of Justice, 347.

3.1 The Moral Imperative of Justice

To protect individuals from dying prematurely is one of the reasons for making medical care available and a major reason it is so very important. That is certainly what is implied by the way in which two of Nussbaum's human capabilities are stated: With respect to the capability entitled "life," she explains that it is the entitlement of "being able to live to the end of a human life of normal length" and "not dying prematurely"; with respect to "bodily health," it is the entitlement of "being able to have good health."[64] Including these concerns has the effect of moving her in the same direction as the two demands of justice we identified as the duties to nurture and to rescue. However, since she does not explicitly acknowledge nurture and rescue as moral demands of justice, Nussbaum is not confronted by the question they raise with respect to her call for a minimum of medical care for all: How does one, how can she, justify inequalities with regard to meeting these duties when life and health are at stake?

Nussbaum's depiction of the capability of "life" includes what is at best an ambiguity, at worst a serious flaw. The overall idea is that individuals should receive the support they need to enjoy a normal lifespan. Adding to, or explicating that idea, is the following assertion: "not dying prematurely, or before one's life is so reduced as to be not worth living."[65] The most benign reading would be to interpret her as saying that people are entitled to the kind of interventions that are necessary to ward off debilitating illnesses as long as possible and to do so in a timely fashion. Another reading might be that the entitlement for the support of an individual's life is no longer necessary once that life is "not worth living." Regardless of what Nussbaum had in mind, the reference to a life not worth living is highly problematic to say the least, especially in the context of specific admonitions intended to guide government policies. One would hope that constitutional law in the United States would continue to uphold an unqualified interest in protecting human life.

In *Washington v. Glucksberg*, the late Chief Justice Rehnquist defended the unqualified protection of human life in several of his assertions: This protection is continuous and does not diminish in strength or weight since it does not depend "on the medical condition and the wishes of the person whose life is at stake"; states "may properly decline to make judgments about the 'quality' of life that a particular individual may enjoy"; "all persons' lives from beginning to end, are under the full protection of the law."[66] Such unqualified protection of human life is vital to assuring protection for the lives of individuals and groups. This was certainly the clearly articulated view expressed by the Federal Constitutional Court of West Germany when it maintained the following: "The express incorporation into the Basic law of the self-evident right to life—in contrast to the Weimar Constitution—may be explained principally as a reaction to the 'destruction of life unworthy of life,' to the 'final solution' and 'liquidations' which were carried out by the National Socialist Regime as measures of state."[67] This Court's unwillingness to countenance quality

---

[64] Ibid., 76.
[65] Ibid.
[66] Washington v. Glucksberg, 117 S. Ct. 2258 (1997), 2272.
[67] Federal Constitutional Court of West Germany, in Ethics in Medicine, Stanley Joel Reiser, Arthur J. Dyck, and William J. Curran, eds. (Cambridge, M.A., 1977), 417.

of life judgments is very explicit in its declaration that: "Where human life exists, human dignity is present to it; it is not decisive that the bearer of this dignity himself be conscious of it and personally know how to preserve it."[68]

We are not at all suggesting that Nussbaum in any way condones the atrocities of the Nazi regime. Rather we are suggesting that the worth of human lives be held to be equal and beyond calculation, and that the term "dignity" is not without its ambiguities and contested usages. Since Nussbaum contends that her list of capabilities specify what it means to live a life of dignity, introducing the idea that a life of a certain kind may not be dignified or otherwise worth living, if it were to be adopted would undermine, certainly not support the moral rationale for the role of psychiatrists, other professionals, as well as various private and public services, in assisting individuals to affirm their worth and not make decisions on the basis of a negative view of themselves. The duties to rescue and care for people should not rest or depend upon how the worth of a life is being assessed by the individual being offered assistance or by anyone else. The worth of an individual should not be questioned.

Medical care as a demand of justice has its anchor in the fundamental role of care in Nussbaum's conception of justice. Nussbaum stresses that all of us are at various stages and times in our lives in need of care, and it is a primary need in times of "acute or asymmetrical dependency."[69] Care should not be thought about as one capability among others. Rather it provides support for all the capabilities Nussbaum lists. Germane to the need for medical care is that: "Good care for dependents, whether children, elderly, ill, or disabled focuses on support for capabilities of life, health, and bodily integrity."[70] Care, for Nussbaum, functions much like nurture as we have characterized it. However, it is introduced in a rather *ad hoc* way, certainly without clarifying whether care, as nurture is for us, a moral demand of justice in its own right. She speaks of care as *supporting* capabilities, the specified goals of justice and does not identify care as one of the goals of justice. In fact, she explicitly rejects that way of thinking about care.

Her conception of care differs from our conception of nurture in yet another significant aspect. We have depicted nurture not only as a strict requirement of justice but also an expression of our empathic emotions which enables us to perceive when care is needed, and at the same time motivate us to be compassionate toward those who need care. Nussbaum is aware that the goals of justice as she has stated them will not be met unless people are motivated to be just. She singles out what she refers to as moral sentiments, identifying among these sympathy, compassion, and benevolence.[71] Although she is aware of neuroscientific research on empathy and cites one major work that contains evidence of people's natural inclination to react compassionately toward those in distress, she emphasizes the necessity to educate

---

[68] Ibid, 419.
[69] Nussbaum, Frontiers of Justice, 168.
[70] Ibid.
[71] Ibid., 408–414.

3.1 The Moral Imperative of Justice

people to be motivated in that way, and does not link her discussion of the need for compassion to her discussion of care.[72] Thus she is left without a natural, motivational and cognitive basis for meeting the goals of justice by means of care for one another.

As noted previously, Nussbaum does not invoke finite resources as rendering any of the goals of justice unobtainable. She rejects what she refers to as traditional moralities that distinguish between positive and negative duties, maintaining as they do that there is a strict duty not to harm others but no strict duty to stop hunger or diseases or to give money to attain their cessation.[73] The capabilities approach rejects this positive/negative distinction as well as the underlying distinction between matters of justice and material aid. As she asserts,

> All the human capabilities cost money to support. This just as true of protecting property and personal security as it is of health care, just as true of the political and civil liberties as it is of providing adequate shelter.[74]

To protect capabilities, the state has affirmative tasks connected with all of them and they all cost money, usually raised by means of taxation that is partially redistributive. In contrast to notions of ownership that claim that people own the unequal amounts they possess, she sides with those from Grotius to Mill who have held that, "the part of a person's holdings that is needed to support members of society (or world, in the case of Grotius) are actually owned by the people who need them, not by the people who are holding on to them."[75]

With a view like this, one would think that Nussbaum would take note of the longstanding tradition of physicians sharing their resources with those in need of both medical and monetary assistance, and obtaining the wherewithal to do so by pegging their fees to a patient's ability to pay. That way, they could also provide free care to the very destitute. However, Nussbaum does not mention this medical ethic and practice. Even more puzzling is that she does not insist that when it comes to medical care, everyone should be treated equally, receiving the care deemed essential to meet each patient's needs. Instead, she calls for a minimum adequate level of care. As it turns out, then, Nussbaum, like Sen, is vulnerable to the argument by Daniels that his Rawlsian theory and the capabilities approach come to the same conclusion: Everyone should be guaranteed some adequate level of medical care, the adequacy of which is to be determined by public policy.

What reasons does Nussbaum have for ending up with a policy on the provision of medical care essentially indistinguishable from that of Daniels with respect to the nature and extent of such care governments, as a matter of justice, are obligated to finance? She does not, like Daniels, cite finite resources and competition among

---

[72] Ibid. Note that Nussbaum does not attend to the existing natural expressions of the empathic emotions that make individual and communal life possible. Care, nurture are key natural expressions of empathy; that she overlooks.

[73] Ibid., 372.

[74] Ibid.

[75] Ibid., 372–373.

governmental obligations as reasons that governmental support for medical care should be limited to less than full coverage for all. One reason she offers for limiting the entitlement to medical care is not a moral consideration *per se*. In her opinion, it is not realistic to expect that societies can achieve an "overlapping consensus," an expression borrowed from Rawls, to approve of more than a decent, or possibly high minimum of medical care for all citizens.[76] Of course, she has no way of knowing whether the opinion is correct unless she, or someone else, puts it to the test by seeking to build consensus around a policy of providing for all the medical needs of individuals. Our concept of justice offers a rationale for such a policy, a policy that is not new. Since Hippocrates, the traditional medical ethic has urged such a policy. It has influenced the practice of medicine and still does. The idea that justice demands it is by no means dead, but it is being buried by many circumstances confronting medical practitioners today, and by philosophical theorizing of the sort we are examining and criticizing in this chapter.

Nussbaum also explicitly argues that justice does not demand anything more than a high minimum of medical care for citizens. Her basic premise is that capabilities are entitlements that set the goals necessary to achieve the equal dignity of human beings. Some policies designed to attain these goals can only succeed, and hence be just, if the capabilities in question are fully, equally accorded to all citizens. These include all the political, religious, and civil liberties. Of these, Nussbaum says, they "can be adequately secured if they are *equally* secured."[77] Any group of people granted less than equality with respect to voting rights or religious liberty would put them "in a position of subordination and indignity vis-à-vis others" and thus "fail to recognize their equal human dignity."[78] At the same time, there are other capabilities for which the goals do not require strict equality with respect to what governments are morally obligated to provide. Housing is one example she discusses, concluding that the housing that should be constitutionally guaranteed need not be equal. Something like the size of a house above a certain threshold is not intrinsically related to equal dignity. Similarly, in provisions for education, at least at the primary and secondary level, adequacy should be the norm, and it is only gross inequalities that clearly constitute a violation of equal dignity to be overcome by setting adequacy at some level she refers to as a "high minimum." Then Nussbaum simply asserts that the same standard of justice applies to "basic essential health care."[79] For this assertion, she offers no argument. What should happen to higher education, nonessential health care, and indeed, with regard to what constitutes adequacy, the level to which people are to be treated equally, she leaves to public deliberation within the various nations of the world.

Nussbaum does not tell us what it is about medical care that justifies inequalities. Nor does she indicate why inequalities would fail to affect what she regards as dignity. After all, dignity is, for her, "truly human functioning" and health is a capabil-

---

[76] Ibid., 294–295.
[77] Ibid., 292–293.
[78] Ibid., 293.
[79] Ibid., 293.

3.1 The Moral Imperative of Justice

ity highly prized and ardently sought. She illustrates this abundantly in her sketches of women in India.[80] She sees this as a universal phenomenon. As it turns out, dignity may be too weak a reed to lean on when it comes to what justice demands. At the very least, Nussbaum's use of the term generates some critical ambiguities and unanswered questions. Why wouldn't the denial of medical care that others receive be experienced as an indignity, and why isn't the failure to treat similar cases similarly a fundamental violation of justice? Whereas these questions are not answered by Nussbaum's approach to the demands of justice, we regard these inequalities as clear violations of justice. Our reasons for making this claim were spelled out in our discussion above of the hypothetical case of Jack and Jill introduced by Daniels.

It is difficult to see why Nussbaum falls short of favoring a policy that would insist upon making full medical care for everyone a strict demand of justice, a requirement for realizing a life of dignity and equally so for all. Consider her contention that needing care is an aspect of our dignity.[81] Equal dignity is what justice demands as spelled out in her list of capabilities. How then can she justify anything less than fully meeting everyone's genuine needs for medical care to attain and retain the capabilities of life, bodily health, and bodily integrity, to cite three of the ten capabilities on her list? Consider the relationship Nussbaum depicts between dignity and capabilities: "Dignity is not defined prior to and independently of capabilities."[82] The argument for the capabilities she lists as the goals of justice is that, with respect to each of them, to imagine a life without the capability in question is to imagine a life not worthy of human dignity.[83] With this argument of hers in mind, ask yourself what level of medical care is required to render secure the capability of bodily health, and therefore, your ability to live a life of dignity. Would that level be anything less than the availability of the medical care you and your physician would deem necessary and possible to protect, maintain, and restore your health, and when necessary and possible, to save your life? Would not something less, some minimum of medical care, threaten to diminish, or even in some circumstances, destroy your health, and hence your ability to live a life worthy of dignity? According to Nussbaum's view of dignity and capabilities, she should answer yes, but doesn't.

Although Amartya Sen does not differ from Nussbaum in endorsing a policy of universally available, minimum medical care, he takes a somewhat different route to defending that policy. Like Nussbaum, he views justice as a plural set of moral demands that governments guarantee some adequate level of the capabilities necessary to be able to choose to live "good and worthwhile lives."[84] Instead of resorting to the concept of dignity, however, he uses freedoms as measures by which to assess what makes human lives good and worthwhile. Sen does not draw up a specific list of capabilities essential for securing the freedoms justice demands. Nevertheless,

---

[80] Nussbaum, Women and Human Development.
[81] Nussbaum, Frontiers of Justice, 160 and 356.
[82] Ibid., 162.
[83] Ibid., 78.
[84] Sen, The Idea of Justice, 226.

many of his illustrative, suggested policies refer to capabilities very similar to those found on Nussbaum's list. The capability of health is among these and it has a prominent place in Sen's concern that governments pursue justice.

"Freedom to determine the nature of our lives," is for Sen, "one of the valued aspects of living that we have reason to treasure."[85] Sen offers two reasons why freedom is valuable: (1) The more freedom we have the more opportunity we have to pursue our objectives; (2) "we may attach importance to the process of choice itself."[86] Any theory of ethics, particularly any theory of justice, "has to decide which feature of the world we should concentrate on in judging a society and assessing justice and injustice."[87] Contrasting his capabilities approach to those who use utility or resources to judge individual advantage, Sen advocates that we judge individual advantage "by a person's capability to do things he or she has reason to value."[88]

According to Sen, capability is concerned with the ability to achieve a combination of functions: "The various attainments in human functioning that we may value are very diverse, varying from being well nourished or avoiding premature mortality to taking part in the life of the community and developing the skill to pursue one's work-related plans and ambitions."[89] Sen sees this approach as a "departure from concentrating on the *means* of living to the *actual opportunities* of living."[90] He contrasts his approach with that of Rawls, regarding his as a means-oriented approach, focusing on primary goods. Sen calls these primary goods "all-purpose means" and cites some of them, such as income, wealth, powers, prerogatives of offices and the social basis of self-respect.[91]

What is curious about Sen's criticism of Rawls is that he does not identify or refer to the primary good of opportunity yet Sen is describing opportunities as goals of justice, as a concomitant of capabilities or as what capabilities make possible. It is difficult to see this as a clear departure from Rawls as it is in the case of income and wealth. The difference between "powers" and the ability that make it possible to make choices also need clarification. One can certainly understand why Daniels failed to find significant difference between his approach and that of Sen, particularly as to medical care, the goal of which is to sustain people's normal range of opportunities.

In addition to freedom, Sen recognizes equality as a demand of justice. In fact, he notes that, "every theory of social justice that has received support and advocacy in recent times seems to demand equality of *something*—something that is regarded as particularly important in that theory."[92] Thus, as he observes, the battles on distri-

---

[85] Ibid., 227.
[86] Ibid., 228.
[87] Ibid., 231.
[88] Ibid.
[89] Ibid, 233.
[90] Ibid.
[91] Ibid.
[92] Ibid., 291.

3.1 The Moral Imperative of Justice

butional issues tend to be not about "Why equality?" but about "equality of what?"[93] For Sen, equality of capability is significant, but other considerations can sometimes outweigh such equality. An important reason for this is that capability is only one aspect of freedom, having to do with substantive opportunities and does not adequately address fairness and equity as these pertain to procedures of relevance to what justice demands.[94]

To illustrate his point, Sen constructs a hypothetical example in which requirements for an equitable process, one that insists that different persons be treated similarly in life and death matters, clearly overrides strict equality in the opportunity aspect of freedom. He calls attention to the fact that, given similar medical care, women tend to live longer than men and have lower mortality rates in each age group. If one were concerned exclusively with equality of the capability to live, one could construct an argument on behalf of giving men relatively more medical attention than women to overcome this masculine handicap. That would be in clear violation of what justice requires, namely that different persons be treated similarly in life and death matters, an aspect of equality that Sen concludes is an instance in which equity in how people are treated can override an appeal to observing equality of a particular capability.

Sen's concern for equality and addressing inequalities is underlined in another reference to matters of health and medical care. In the context of criticizing the current predilection to use the gross national product (GNP) and the gross domestic product (GDP) as measures of human well-being, he points out that economic wealth and substantive freedom, though connected, can diverge with respect to matters of health and medical care. He notes that: "Even in terms of being free to live reasonably long lives (free of preventable ailments and other causes of premature mortality)…very rich countries can be comparable to that of developing economies" with respect to the extent of deprivation experienced by certain disadvantaged groups.[95] While he acknowledges that freedom from premature mortality is helped by having a higher income, Sen observes that this freedom depends on other features of a society, such as, "public healthcare, the assurance of medical care, the nature of schooling and education, the extent of social cohesion and harmony, and so on."[96]

Despite Sen's attention to inequalities with respect to health and longevity, and the availability of medical care, the policy he regards as just in providing medical care is always that of publically guaranteeing that there is universal access to *basic* medical care. To cite one example among others, he rejects the use of utilitarian measures, particularly happiness and well-being as measures of substantive freedom, by calling attention to the neglect of "positive freedoms" when such measures are used. Sen cites "basic healthcare" as a prime example of what is neglected by the resort to utilitarian measures such as happiness and well-being.[97]

---

[93] Ibid., 295.
[94] Ibid.
[95] Ibid., 226.
[96] Ibid., 227.
[97] Ibid., 282.

However, this commitment to basic medical care conflicts with Sen's earlier insistence that different persons be treated similarly in life and death matters. After all, people with a high enough income can receive the medical care physicians indicate they need for preventing, curing, or otherwise caring for their illnesses that, without treatment, would undermine their health, cause them suffering, or threaten their lives. Whether basic medical care will offer the same benefits is not specified in any of Sen's references to it. If it did, why speak of *basic care* if what you are advocating is having all of your medically diagnosed, treatable needs, met? Why not expressly call for a policy guaranteeing people's need for medical care as specified by the existing professional standards of what constitutes such a need? In any event, Sen certainly does not differentiate his view from that of Daniels and Nussbaum: What justice demands is some adequate or decent minimum of medical care, left to be determined by processes that are fair. Like Daniels and Nussbaum, Sen does not entertain or attend to the standard of justice in the provision of medical care that we have identified with the traditional medical ethic.

3.2 The Relation Between Justice and Rights

Daniels, Nussbaum, and Sen agree: Rights are important and they feel compelled to consider the sense in which medical care is a right. They note that there are sharp disagreements about how to conceive of rights and whether they are universal. In addition they acknowledge the philosophical skepticism about any sound basis for asserting them. Nevertheless, each of these scholars end up favoring a universal entitlement to a basic minimum of medical care. This basic entitlement is, for each of them, at once a demand of justice and a right that should be guaranteed by governments.

Readers will recall how we depicted the relationship between justice and rights in our first chapter. The demands of justice arise from the empathic emotions we share as human beings. Since we share these demands of justice, these moral requisites of our individual and communal existence, we expect to be treated justly. Since human beings are also capable of injustices, such as neglecting to nurture life or of directly inflicting harms, some of our expectations take on the character of claims. Thus we have laws and institutions, for example, that enforce certain aspects of nurture and certain prohibitions of serious harms. All of that has been described in more detail in the first chapter. At this juncture, we remind our readers of the view we are taking of the relationship between justice and rights. The moral demands of justice identify what we owe one another: That means they identify our moral responsibilities toward others and our rights, namely what others owe us.

Daniels, Nussbaum, and Sen, each in their own way, also portray rights as demands of justice and for justice. However, our analysis of their accounts of rights will reveal how these accounts differ from our own.

3.2.1 Daniels on Rights to Health and Health Care

Daniels clearly affirms a right to health and a right to health-care services that are one of the necessary means to achieving a right to health. He agrees with philosophers who claim that the range of human rights recognized in international law "lack an adequate philosophical foundation."[98] However, he expressly contrasts his approach to "justifying a moral right to health and the positivist legal approach to human rights embodied in international law."[99] He does not consider moral rights as a way to discover what we owe one another with regard to protecting health. Rather, moral rights should emerge from "broader work in political philosophy."[100] (As we know, that "broader work" for Daniel's is that of John Rawls on justice as fairness.) Accordingly, as Daniels goes on to say,

> I used the observation that protecting normal functioning helps to protect fair equality of opportunity to ground my view that the moral right to health and health care is a special instance of a broader right to fair equality of opportunity…If we have a social obligation to ensure equality of opportunity, we have an obligation to promote normal functioning, and our moral right to health and health care is the corollary of those obligations.[101]

Thus we discover our right to health care when we discover what justice demands in that respect.

That sounds as if Daniels shares our view that what you have a right to is what justice demands of us as human beings. However, his understanding of this relationship differs from ours in some rather important ways. He explicitly asserts that his account "fails to provide a foundation for a *human* right to health and health care, one that arises simply because we are human."[102] His justification is tied to the kind of societal circumstances existing and justified by the Rawlsian contract theory of justice. For that reason, Daniels indicates that his justification of a right to health and health care does not establish its universality. For him, a *human* right is universal in that it is "independent of the social relationships that exist in a state or society."[103]

When Daniels speaks of existing social relationships, he does not have in mind the moral relationships that we refer to as moral requisites of individual and communal life. These are universally present in any functioning community to a degree that sustains that community. Furthermore, as we have indicated, these moral relationships have a source in the shared human proclivities and inhibitions that arise from our empathic emotions. As a result, some human relationships are universal, and these include those compassionate relationships that obtain between those who take up the art of healing, and those in need of care or rescue. These relationships express and satisfy a human moral responsibility to care for and rescue those in

---

[98] Daniels, Just Health, 314.
[99] Ibid., 316.
[100] Ibid.
[101] Ibid.
[102] Ibid.
[103] Ibid., 317.

need, and at the same time, that satisfies the natural right to receive such care and rescue. Characterizing rights in this manner derives the content of a right from the content of universal moral responsibilities. Daniels is on a different page, asserting that even when rights are deemed to be universal human rights, "in practice the content of the rights that emerge similarly depend on the conditions in a specific state."[104] For him, then, the content of all rights are context-dependent.

It is certainly true that what specific rights people actually enjoy, and how and to what degree these rights are realized and supported, will in fact vary from society to society, and from state to state. It it is true also that human rights, despite the universality of the demand for them, are in practice violated and/or denied within societies and states, again to various degrees and in different ways. However, there would be no justification for making such judgments if we were unable to know what we universally owe one another as human beings. How else could we assess the degrees to which rights are being actualized or denied in any society or state?

Daniels is not observing a very critical distinction. Saving a human life is a right-making characteristic of actions: As such it is universally and always so. Intentionally killing someone is a wrong-making characteristic of actions: As such, it is universally and always so. Saving a human life is just; intentionally killing an innocent individual is unjust. Specific human actions, however, are almost always morally complex. Hence, it is intentionally killing an *innocent* person that is, as a particular act, always wrong. Thus a police officer may have to shoot someone from a distance to save an innocent baby from an individual poised to strike that infant with a lethal weapon. Also, in certain circumstances, it may be wrong to attempt a rescue because anyone attempting it would die in the effort. Similarly, the interventions of physicians are virtually always complex in this way. Physicians are rightly inhibited from harming anyone, yet they rightly subject their patients to procedures that knowingly put patients at risk of harm or that knowingly inflict harm with the goal of improving the health of such patients or of saving their lives. Yet, if they have their facts straight and have correctly weighed the harms and benefits of their actions, they are acting justly and in accord with the moral rights of their patients.

The content of the moral demands of justice and the rights to be treated justly are not as such context-dependent. The content of a specific act or policy, and the extent to which it is just, are context-dependent. Actions and policies are characterized by goals and means to achieve those goals. Decisions whether to act in a particular way, or to follow a particular policy, are not only based upon the universal demands of justice, our identifiable responsibilities and rights, but also upon a number of non-moral facts. Among other things, these facts include those moral demands of justice that are relevant to the act or policy in question, what consequences may result from that act or policy, and what is possible or impossible to set as a goal or to have the means to achieve the goal being set.

In order, then, to attain justice, we need knowledge not only of what responsibilities and rights can be identified as universal moral demands of justice, but also of the non-moral circumstances in which we are acting. We should not confuse the task

---

[104] Ibid.

of identifying and characterizing the content of universal moral demands, rights and responsibilities, with the task of determining whether a specific action or policy qualifies as just. As already noted, the actions of physicians that can reasonably be considered just, morally responsible, almost always involve carrying out actions that, taken by themselves, are morally wrong, such as those required in carrying out surgical procedures. Therefore, physicians are compelled by law to obtain the informed consent, except in emergencies, of their patients for what would otherwise be an unlawful, morally unjustifiable, assault on their right to be protected from being harmed.

Daniels is well aware of the uncertainties attending decisions about specific acts and policies. However, his unwillingness to consider a right to healthcare as a human right does affect how he thinks such decisions should be made. As he correctly points out: "The general principles of justice for health that we have been discussing are simply too general and too indeterminate to resolve many reasonable disputes about how to allocate resources fairly to meet health needs, and we lack consensus on more fine-grained principles."[105] The principles that Daniels invokes are indeed general and indeterminate. For him, a right to health and health care is "a special instance of a broader right to fair equality of opportunity."[106]

Since the right to healthcare Daniels proposes is a right to some basic minimum of care, and since general principles or appeals to human rights will not in his view indicate what the content of that minimum care should be, he calls for a process of deliberations he labels "accountability for reasonableness."[107] The task for this process is to limit heath care while doing so in a way that people can agree is fair. Accountability for reasonableness is a process of decision-making that satisfies four conditions: (1) The decisions and their rationale must be publically accessible; (2) the rationale for any decision must be reasonable, that is, appeal to evidence, reasons, and principles accepted as relevant by people who are fair minded and disposed to finding mutually justifiable terms of co-operation; (3) there must be in place procedures to allow for appealing and revising decisions; and (4) the process must be subject to voluntary or public regulations that ensure that conditions 1–3 are met.[108]

Laudatory as these conditions are, they fall short of being just or fair. Missing from the guidance of these deliberations are the human rights that should be considered if the provision of medical care is to satisfy the moral imperatives of justice. These human rights reflect what justice demands. They are quite "fine-grained" and supported by the empirical and rational evidence we have cited in our discussion of them as moral requisites of individual and communal life. In contrast, then, to Daniels, we insist upon deliberations and decisions that recognize a human right to medical care that meets all needs deemed morally and medically possible and necessary to save lives, prevent, ameliorate, or cure illnesses and relieve the suffering

---

[105] Ibid., 103.
[106] Ibid., 316.
[107] Ibid., 118.
[108] Ibid, 118–119.

of those with such needs. Daniels chooses instead to limit such care on financial rather than on purely medical grounds when limiting care is in the patient's best interests. This position of Daniels constitutes an acceptance of unfairness in that people of sufficient means will be able to have their medical needs met fully, while those dependent upon financial assistance will have to be content with some minimum of medical care. Daniels clearly does not adhere to the traditional medical ethic in what he alleges is a fair process for making decisions about limiting medical care. Such decisions should be based upon benefiting patients, not upon depriving them of medically justifiable interventions: Patients have a right to receive such care; physicians have a right to dispense it.

3.2.2 Capabilities and Rights: Nussbaum

In depicting the relation between rights and capabilities, Nussbaum offers us two somewhat differing accounts. One in the year 2000 reads as follows:

> Capabilities as I conceive them have a very close relationship to human rights, as understood in contemporary international discussions. In effect they cover the terrain covered by both the so-called first generation rights (political and civil liberties) and the so-called second generation rights (economic and social rights). And they play a similar role, providing the philosophical underpinning for basic constitutional principles.[109]

Given this description of the relation between rights and capabilities, Nussbaum is led to ask what it is that the language of capabilities adds to the well-established language of rights. First of all, how to conceive of rights is in dispute. Some of the disagreements she cites are over: The basis of rights claims; whether rights are pre-political or artifacts of laws and institutions; whether rights belong only to individuals or to groups as well; what relationship exists between rights and duties; and "what rights are to be understood as rights to."[110] The advantage Nussbaum attributes to her portrayal of capabilities is that "of taking clear positions on these disputed issues, while stating clearly what the motivating concerns are and what the goal is."[111]

A second reason to supplement rights language with the language of capabilities is that capabilities provide a "benchmark" for what it takes to secure a right for someone.[112] For example, if women in many nations are to have more than a nominal right to political participation, they will also need to have support for capabilities necessary to being politically active. Rights, Nussbaum suggests, may best be thought of as "combined capabilities."[113]

---

[109] Nussbaum, Women and Human Development, 97.
[110] Ibid.
[111] Ibid.
[112] Ibid., 98.
[113] Ibid.

3.2 The Relation Between Justice and Rights

There is yet another advantage Nussbaum attributes to the language of capabilities over against the language of rights: "it is not strongly linked to one particular cultural and historical tradition, as the language of rights is believed to be."[114] Although Nussbaum does not think that rights are exclusively Western, she favors the language of capabilities because it allows us to bypass any debate about that.

Despite these advantages of the language of capabilities over against the language of rights, Nussbaum emphatically wishes to retain the language of rights. In her view, rights language has four important roles to play in public discourse. To begin with, asserting rights serve to remind us that people "have justified and urgent claims to certain types of treatment no matter what the world around them has done about that."[115] For example, individuals have a right to have the basic political liberties secured for them by their government. Secondly, rights language emphasizes the importance of the abilities and functions appearing on the capabilities list and backs that up "by a sense of the justified claims that all humans have to such things by virtue of being human."[116] Third, rights language is valuable because of its commitment to people's choice and autonomy. The language of capabilities leaves room for choice but some approaches to capabilities in the Aristotelian tradition give insufficient attention to liberty. Finally, in the area of disparate analyses of rights talk, "language of rights preserves a sense of the terrain of agreement" while deliberations continue "about the proper types of analysis…"[117]

In the year 2000, then, Nussbaum distinguishes the language of rights from the language of capabilities. However, it is difficult to see the significant difference between political participation such as voting being called a capability and this same political participation being called a right, when in both instances this freedom is to be guaranteed by governments and enshrined in their constitutions. Nussbaum appears to have come to that same conclusion when, in 2006, she declares that "the capabilities approach is fully universal" and it is "one species of a human rights approach."[118] Nussbaum does not clarify what it means to refer to her capabilities approach as a "species" of a human rights approach. Actually, or so we will contend, what Nussbaum has done is to develop a particular theory of human rights.

A first clue to understanding Nussbaum as offering a distinctive theory of human rights is her statement, similar to the one cited earlier, that:

[C]apabilities cover the terrain occupied by both the so-called first generation rights (political and civil liberties) and the so-called secondary rights (economic and social rights). And they play a similar role, providing an account of extremely important entitlements that can be used as a basis both for constitutional thought within a nation and for thinking about international justice.[119]

---

[114] Ibid., 99.
[115] Ibid., 100.
[116] Ibid.
[117] Ibid, 101.
[118] Nussbaum, Frontiers of Justice, 284.
[119] Ibid.

Notice that the capabilities not only cover the same territory covered by human rights, but also constitute an account of important fundamental entitlements. Now, entitlements and what becomes constitutional law, specify societal obligations that should, as a matter of justice, be guaranteed by governments. What is being called for as a matter of justice, then, are surely rights, that is, claims individuals have upon their governments and societies to meet certain kinds of responsibilities toward them. Furthermore, these entitlements, like human rights as Nussbaum has described them, are prepolitical: "Thus," as she says, "a nation that has not recognized these entitlements is unjust."[120]

Still, Nussbaum persists in distinguishing the language of rights from the language of capabilities, which in her words, "gives important precision and supplementation to the language of rights."[121] Or again, "thinking in terms of capability gives us a benchmark as we think about what it really is to secure a right to someone."[122] What she could argue is that she has given us a specific and precise account of rights by tying them to capabilities.

For her justification of each of the capabilities, Nussbaum relies upon the concept of human dignity. To generate her list of capabilities, she begins "with a conception of the dignity of the human being and of a life worthy of that dignity—a life that has available in it "truly human functioning'..."[123] She then uses this idea of dignity to justify a list of ten capabilities as "central requirements of a life of dignity."[124] Each of these capabilities can be justified, if by imagining a life without the capability in question, we can conclude "that such a life is not a life worthy of human dignity."[125] Though she describes such an argument for each capability as intuitive and discursive, she is convinced that broad cross-cultural agreement can be achieved for both the list and the process. She notes also that "rights have often been linked in a similar way to the idea of human dignity."[126] We need to say more now about Nussbaum's conception of human dignity.

Nussbaum indicates that human dignity has to be understood in terms of equality: "it is the *equal dignity* of human beings that demands recognition."[127] She then evaluates what policies are required as a matter of justice. In effect, equal dignity serves as the right-making characteristic of the various policies which have as their goal some level of the capabilities called for by a life of equal dignity for all members of a given society, to be guaranteed by each society's government if they are to be considered just.

Nussbaum has, in our view, unwittingly developed a theory of rights: Equal dignity is *the universal right*; capabilities guided by this right are the rights being con-

---

[120] Ibid., 285.
[121] Ibid, 285.
[122] Ibid, 287.
[123] Ibid., 74.
[124] Ibid., 75.
[125] Ibid., 78.
[126] Ibid.
[127] Ibid., 292.

cretely suggested by setting goals that meet the standard set by equal dignity. This appears to be a plausible way of providing a road map, what she calls benchmarks "as we think about what it really is to secure a right to someone."[128]

Unfortunately, as we have already discovered in our earlier discussion of Nussbaum, her appeal to equal dignity is flawed in two very important ways. First, dignity is subject to more than one interpretation, at least as she presents it; the most troubling interpretation is that it raises questions about the equal worth of life. The other flaw is that she maintains that equal dignity requires strict equality of certain liberties, such as the right of all individual citizens to vote, while allowing for inequalities in support for certain other capabilities. Given our focus, the most notable one is that of accepting some basic minimum of medical care as sufficient to satisfy what it means to be guaranteed equal dignity. Lacking a generic concept of justice, Nussbaum has no basis for strict justice in the provision of medical care, namely meeting those needs assessed by medical standards as necessary if one's health, well-being, and life are to be sustained, regardless of one's ability to pay for such services. She does not explicitly refer to the traditional Hippocratic ethic nor to any of the reasons for its existence. By our understanding of equal dignity, she would have a reason to advocate that traditional medical ethic. However, on her understanding of equal dignity and of justice as such, she has no such grounds. Indeed, she lacks a justification for the policy on medical care she adopts.

3.2.3　Rights as Freedoms: Sen

Sen regards rights in much the same way as Nussbaum. First: "The notion of human rights builds on our shared humanity."[129] Secondly, "human rights are really strong ethical pronouncements as to what should be done"; they are not claims that are "already established *legal* rights, enshrined through legislation or common law."[130] Rather, as ethical imperatives, they "can serve as grounds for legislation."[131] As we have noted, Nussbaum refers to this characteristic of human rights as "prepolitical."

Whereas Nussbaum assesses what is to count as a right by asking whether it is demanded to assure individuals equal dignity, Sen appeal to freedom for this purpose. For Sen, "declarations of human rights are…ethical affirmations of the need to pay appropriate attention to the significance of freedoms incorporated in the formulations of human rights…"[132] To decide what qualifies as a human right, it is necessary to reflect upon whether a given freedom is one that merits our serious

---

[128] Ibid., 287.
[129] Sen, The Idea of Justice, 143.
[130] Ibid., 357–358.
[131] Ibid., 363
[132] Ibid., 366.

attention and response. Like Nussbaum, Sen introduces us to freedoms as rights and others as lacking the requisite significance attached to rights claims.

Sen has us imagine the following five freedoms, all of which are of considerable importance to a fictitious person he names Rehana:

(1) Rehana's freedom not to be assaulted;
(2) her freedom to be guaranteed some basic medical attention for a serious health problem;
(3) her freedom not to be called up regularly and at odd hours by her neighbors whom she detests;
(4) her freedom to achieve tranquility, which is important for Rehana's good life;
(5) her freedom from fear of some kind of detrimental action by others (going beyond the freedom from the detrimental actions themselves).[133]

With respect to the first, the freedom not to be assaulted, and the second, the freedom to receive basic medical attention, Sen simply asserts the plausibility of both to be regarded as human rights. This is interesting because the right to basic medical care has already been defended by Sen as a matter of justice, and for that reason an entitlement governments should guarantee. One would think that Sen would acknowledge that and explicitly recognize this relationship that exists for him between justice and rights. However, he fails to do so. As for the first freedom, one suspects the obvious injustice of being assaulted without provocation is the reason Sen endorses freedom from being assaulted as an obvious candidate for consideration as a human right. Sen, however, needs to qualify this right. Rehana would forfeit her right not to be assaulted in any situation in which she is assaulting someone in a seriously harmful way.

Sen does not regard the freedom to be free of disturbance from unloved neighbors as having enough "social relevance to qualify as a human right."[134] It is curious that he does not raise the question as to whether these disturbances could be serious enough to constitute harassment. If he were working with our generic conception of justice, and explicitly relating moral rights to the responsibilities that actualize them, he would have reasons to consider circumstances that would rise to the level of an injustice, and in the most serious cases would involve legal redress for harms suffered. This argument applies just as well to the freedom to achieve tranquility which Sen also thinks is not a suitable candidate for qualifying as a human right. Sen argues that this freedom is "too inward-looking and beyond the effective reach of social policies" to be a human right.[135] That would surely depend on the circumstances.

In his discussion of the fear of negative action by others, Sen does acknowledge that whether it is a human right does depend upon certain contingencies, particularly "the basis of that fear, and how that can be removed."[136] Thus, for example, he

---
[133] Ibid., 367–368.
[134] Ibid., 368.
[135] Ibid.
[136] Ibid.

makes a case on behalf of a right to warranted medical attention aimed at removing certain kinds of fears. He argues as well that: "There can even be a reasonable case for placing elimination of the fear of terrorism within the concerns of human rights, even if the fears were stronger than probability statistics would justify."[137]

From the perspective of our generic concept of justice the example of terrorism would come under the responsibility to protect life, a responsibility that actualizes the right to be protected. This responsibility and this right is at the heart of what governments are obligated to do on behalf of security. In short, Sen is not providing explicitly moral reasons for deciding that a given freedom is a human right. That results in considerable unnecessary ambiguity and arbitrariness in his attempts to determine the importance of any particular freedom. That is surprising, since Sen had already affirmed the freedom to have access to basic medical care on the basis of what justice demands and would have good reason to ask what justice would demand with respect to all five freedoms reflected upon in his thought experiment.

The most serious question we have for Sen, however, will not be answered by pointing out how he ought to relate his conception of justice to his conception of rights. As we have indicated in our earlier analysis of Sen's view of the moral demands of justice, he does not think that justice demands that individuals have a right to the full measure of medical care called for by the ideal of justice inherent in the traditional medical ethic. Indeed, in his thought experiment reflecting upon what freedoms qualify as rights, Sen formulates the freedom to be guaranteed medical care as a freedom to receive some *basic medical attention*. Expressing this freedom in that manner is a clear indication that the right he is advocating will fall short of a right to have each individual's medical needs fully attended to, as fully as would be the case for those who have the means to receive all the physicians determine is in their best interest. In short, Sen's version of this right is one that bypasses and does not comply with the Hippocratic ethic. As we have argued, the Hippocratic ethic entails moral demands that can only be perceived from the vantage point of a generic conception of justice. From this same vantage point, one can perceive and articulate as well certain rights that play a vital role in how medical care is and should be distributed. Daniels, Nussbaum, and Sen all fail to recognize or attend to these rights and responsibilities. The failure of Sen, Nussbaum, and Daniels even to consider a human right to have one's need for medical care fully met, stems also from certain inadequacies in their theories of moral cognition.

---

[137] Ibid., 369.

3.3 Inadequate Accounts of Moral Cognition: Sen, Nussbaum, and Daniels

Our previous chapter was devoted to identifying the cognitive bases we have for deciding whether any particular act or policy can justifiably be considered just. To justify a given action or policy requires us: (1) To obtain as much relevant factual knowledge as possible; (2) to vividly imagine how others will be affected by the action or policy in question; and (3) to judge that action or policy from an impartial perspective. With these cognitive processes in mind, we will assess how Sen, Nussbaum, and Daniels go about seeking to justify their view that justice is sufficiently realized when everyone is guaranteed what they refer to as <u>basic</u> medical care.

3.3.1 Moral Cognition in Amartya Sen

Sen is certainly on the right track when he appeals to an impartial perspective as the overall cognitive basis for discerning what constitutes justice in matters of policy. Sen rightly argues that, in the quest for objectivity, ethical judgments need to be impartial: "The reasoning, that is sought in analyzing the requirements of justice will incorporate some basic demands of impartiality, which are integral parts of justice and injustice."[138] For his understanding of impartiality Sen draws upon Adam Smith's conception of an ideal impartial spectator.

Smith's notion of impartiality appeals to Sen because it stresses the need to inform one's ethical reasoning by invoking "a wide variety of viewpoints and outlooks based on diverse experiences from far and near" and not simply on the "same kind of experiences, prejudices, and convictions about what is reasonable and what is not, and the same beliefs about what is feasible and what is not."[139] If objectivity with respect to one's ethical and political judgements is to be realized, reasoning must take into account different perspectives.

For Sen, as for Smith, impartiality demands not only objectivity but also inclusiveness. Justice, Sen maintains, "has a universal reach, rather than being applicable to the problems and predicaments of some people but not of others…The universality of inclusion…is, in fact, an integral part of justice."[140] Sen understands Smith's concept of impartiality as one that views the reach of justice as going beyond any fixed neighborhood. This means that our concern for justice and rectifying injustice obligates us on a global scale; no one, and no people, are excluded from the moral imperatives of justice. Impartiality therefore provides a justification for universal access to medical care as a universal right.

---

[138] Ibid., 42

[139] Ibid. 45.

[140] Ibid, 117. Although the reference of this juncture is to Wollencroft, the view is also that of Adam Smith.

3.3 Inadequate Accounts of Moral Cognition: Sen, Nussbaum, and Daniels

Sen holds that this right to universal access is also a right "to receive basic medical attention."[141] Now Sen never spells out what basic medical care includes and what it excludes from meeting everyone's medical needs fully. Nor does he explain why, in the case of medical needs, he is not asserting strict impartiality in meeting them by insisting that individuals are to be treated similarly in similar circumstances: Surely doing so is a matter of justice, justifiable on grounds of impartiality in provision of medical care.

Sen recognizes equality as a moral imperative of justice. Why not in the right to medical care? What he does is omit from the process of justifying specific actions and policies the requirement that our actions and policies be judged by vividly imagining how they affect everyone. For this to be as close to being accurate and inclusive as humanly possible, we have to engage our emphatic emotions and apply them universally. When we do so, applying them to ourselves as well, we can vividly imagine what it would be like to be denied a particular, necessary, and available medical treatment for lack of the finances to receive it while at the same time those who have the finances receive the same care. Inequality of this kind would almost certainly strike us as unjust, particularly in instances that would result in suffering or an untimely, avoidable death. As noted in Chap. 1, stories of that sort led the U.S. Congress to allocate funds to cover dialysis for those with insufficient resources to pay for being treated.

Although Sen regards sympathy as well as reason as shedding light on thinking correctly about what is just, it does not lead him in any explicit way to call for strict equality in having our medical needs met. Had he tied the Golden Rule to our ability to sympathize, he would have had a cognitive basis for rejecting as unjust to deny medical care to him and anyone else necessary and available medical care for lack of finances, care that others can and do receive.

Sen might well object to any suggestion that his advocacy of basic health care for everyone would countenance a failure to guarantee lifesaving care for all. Indeed, in a discussion of freedom from premature mortality, a capability that should be promoted by governments in the name of justice, he includes the assurance of medical care as one of the means for seeking to achieve freedom from preventable illnesses and other causes of premature mortality.[142] However, Sen does not explicitly argue for guaranteeing everyone all the available medical interventions that are deemed essential by existing medical standards, and doing so equally for those who can pay to make that possible and for those who cannot. Furthermore, Sen has not identified vividly imaging how everyone is affected by a given action or policy as a necessary ingredient for achieving cognitive justification for accepting or rejecting that action or policy; nor has he identified the natural source of the motivation and perceptivity supplied by our empathic emotions to be willing and able to do this, and to do so through the prism of the Golden Rule. Viewing medical care through an empathically motivated and informed cognitive prism, as we have contended, takes us beyond an appeal to basic medical care, whatever that entails, to medical care that

---

[141] Ibid. 368.
[142] Ibid., 226–227.

fully meets everyone's needs for it. Empathy is the fuel and cognitive source for the practice of the Hippocratic ethic.

3.3.2 Moral Cognition in Martha Nussbaum

As previously indicated in this chapter, Nussbaum also makes basic medical care a goal of justice, one to be governmentally guaranteed. She regards basic medical care as necessary to meet certain of the entitlements all individuals should be assured: "A nation that has not recognized these entitlements is to that extent unjust."[143] Among these entitlements are the capabilities of life, health, and bodily integrity. To actualize these capabilities, some measure of health care is required.

Nussbaum grounds her version of the capabilities, to which all human beings are entitled as a matter of justice, upon her "conception of the dignity of the human being."[144] She uses the appeal to dignity as her justification for considering her list of capabilities as entitlements, and thus demands of justice, in the following way:

> The basic idea is that with regard to each of these, we can argue, by imaging a life without the capability in question that such a life is not worthy of human dignity. The argument in each case is based on imaging a form of life; it is intuitive and discursive. Nevertheless, I believe that the process, and the list, can gather broad cross-cultural agreement, similar to the agreements that have been reached concerning basic human rights.[145]

Nussbaum regards her capabilities approach as a "species of a human rights approach" and takes the view that "human rights have often been lined in a similar way to the idea of human dignity."[146]

This appeal to imaging whether a particular policy provides individuals the opportunity to live a life of dignity by guaranteeing individuals a specific capability such as health is akin to the cognitive process we calling "vividly imagining how everyone is affected by a particular policy." However imaging whether a given action or policy supports human dignity is quite different from imagining whether the moral demands of justice, such as saving lives and nurture (which includes the prevention, care, and cure of illnesses), are met by a given action or policy. Justice also includes equality in meeting these moral imperatives. As it turns out, Nussbaum's appeal to dignity fails to satisfy this requirement.

Nussbaum professes to affirm equality as a requisite for achieving a life of dignity. As she says,

---

[143] Nussbuam, Frontiers of Justice, 285.
[144] Ibid., 74.
[145] Ibid., 78.
[146] Ibid.

Equality of capability is an essential social goal where its absence would be connected with a deficit in dignity and respect...it is the equal dignity of human beings that demands recognition.[147]

However, whereas Nussbaum argues that "all the political, religious, and civil liberties can be <u>adequately</u> secured only if they are <u>equally secured</u>," she pegs adequacy for securing health care as requiring "something close to equality or at least a very high minimum."[148] This is puzzling! Surely denying me a necessary medical service that others with sufficient means obtain is a failure to respect my dignity. In any event, I am being treated unjustly, that is, unequally when viewed from an impartial cognitive perspective. Nussbaum does not evoke impartiality.

How does Nussbaum defend this departure from strict equality in what governments should support as a matter of pursuing justice? She argues that a policy that insists upon equality in supporting all of the capabilities is not likely to achieve an "overlapping consensus."[149] For this claim, Nussbaum does not cite any evidence except to note that "people differ in their comprehensive ethical and religious doctrines."[150] However they may differ, all the major religions affirm the Golden Rule. When the Golden Rule is impartially guided and driven by our empathic emotions towards ourselves and others, it would seem to be beyond the pale that we could justify that anyone be denied medically indicated care that we would not wish to have denied to us.

From the perspective of impartiality and our empathically derived ability to imagine how all of us as human beings would be affected by a given action or policy, Nussbaum's attempted justification of inequalities in the availability of medical care is unconvincing. She needs to examine more concretely the reasons people go for medical assistance, and the reasons assistance is granted them by physicians and other caregivers. Policies should accord with justice in its fullest, generic sense, and should be formulated on the basis of the fullest possible factual information, with empathy for all affected, and from an impartial perspective that calls for objectivity, inclusiveness, and equality. Nussbaum falls short of meeting these cognitive criteria for morally justifying our actions and policies.

3.3.3 Moral Cognition in Norman Daniels

Unlike Sen and Nussbaum, Daniels' case for basic medical care is that there are not sufficient resources to do more if society's resources are to be distributed justly. To justify limiting medical care, Daniels depends upon what he considers to be a fair

---

[147] Ibid., 293.
[148] Ibid., 294.
[149] Ibid., 295.
[150] Ibid.

process of reasoning. This process he refers to as "accountability for reasonableness."[151]

Thera are four conditions that characterize the process of accountability for reasonableness. All of them are essential for justifying decisions to set limits to medical care:

1. Publicity condition: Decisions…and their rationales must be publicly accessible.
2. Relevance condition: The rationale for limit-setting decisions should aim to provide a reasonable explanation of how the organization seeks to provide "value for money" in meeting the varied health needs of a defined population under reasonable resource constraints. Specifically, a rationale will be "reasonable" if it appeals to evidence, reasons and principles that are accepted as relevant by ("fair minded") people who are disposed to finding mutually justifiable terms for cooperation. Where possible, the relevance of reasons should be vetted by stakeholders in these decisions….
3. Revision and appeals condition: There must be mechanisms for challenge and dispute resolution regarding limit-setting decisions, and…opportunities for revision and improvements for policies in the light of new evidence or arguments.
4. Regulative condition: There is either voluntary or public regulation of the process to ensure conditions 1–3 are met.[152]

For Daniels, these four conditions are the norms of procedural justice. Such procedural norms are required because, by itself, the more general principle (opportunity) does not resolve disputes about how medical care is to be allocated in a just way. For that, a "fair process is needed."[153] The overall purpose of this fair process is to obtain policies as close to the ideal of justice as possible: Its immediate purpose is to "seek terms of cooperation that rests on justifications acceptable to all." Daniels expects these four conditions to be realized and applied by the personnel of governmental agencies or health insurance companies making decisions about the limits to coverage of medical care they think they should approve. Daniels argues that inequalities can be just, though only if decisions regarding them are made under the four conditions he has specified for using fair procedures to formulate justifications acceptable to all for any limit put upon coverage of medical care.

To asses these conditions properly, it is important to be aware of how they interact with one another. Only then can one understand why Daniels insists that procedural justice is only attained when all four conditions are met.

Consider publicity, the first condition of accountability for reasonableness. This calls for public access to the decisions and rationales that would directly or indirectly set limits upon meeting health needs. The reasons for a decision should be judged for their relevance, the second condition, and where possible, stakeholders in these decisions should have the opportunity to vet the reasons being offered. Both

---
[151] Daniels, Just Health, 112.
[152] Ibid., 118–119.
[153] Ibid., 110.

3.3 Inadequate Accounts of Moral Cognition: Sen, Nussbaum, and Daniels

of these conditions share the kind of information that helps make the third condition possible and meaningful, namely that of providing mechanisms for challenging and disputing limit-setting decisions, and for revising and improving policies in conformity with new evidence or arguments. Then, to assure that those considerations take place, the fourth regulative provision is necessary to make certain that the other three conditions are in fact carried out. That could require legal enforcement.

In effect, Daniels has described processes that address the need for a conscientious pursuit of facts, and that includes reasoning and reasons being generated in decision-making process that will serve to justify policy decisions. Conceived in this way, one can certainly affirm such a cognitive process. It should, however, be stated in its more comprehensive sense as constituting only one of the cognitive processes necessary to justify moral decisions and morality of policies in general.

Daniels requires something more than fact-finding if decisions are to be justifiably just. Those making these decisions have to be "fair-minded." That means they are individuals "disposed to finding mutually justifiable terms of cooperation."[154] Specifying that one has to be a fair-minded individual in order to able to discover reasonable justifications for one's moral decisions is in effect, albeit implicitly, an appeal to an ideal moral judge, one who can make moral judgments from an impartial perspective, that is to say, from a perspective that is as free of bias as possible. The quest for an impartial perspective is underlined by Daniels when he stipulates that the reasons for the decisions being made "should be vetting by stakeholders in these decisions…"[155]

Daniels takes note of another cognitive process that fair-minded individuals should engage in for the sake of morally justifying their policy decisions. There are occasions, Daniels asserts, when disadvantage someone. However, in the event that a decision disadvantages someone (and others like him) more than anyone needs to be disadvantaged under available alternatives then this is a reason that all should consider relevant for questioning such a decision.[156] Were one to approve of a decision with such a result, it would be "the basis of a complaint that each person would want to be able to make if he or she was the person so severely disadvantaged."[157] On this issue, Daniels has in effect evoked the cognitive process of vividly imagining how everyone, oneself included, is affected by an action or policy as a process that should characterize fair-minded individuals and, as we have contended, an ideal moral judge. Again, Daniels would strengthen his case for what constitutes procedural justice if he would explicitly highlight this single appeal of his to imagining how everyone is affected by a policy and recognize that it is a cognitive process that should be used to justify any and all decisions that claim that a given act or policy is just.

Overall, Daniels has given us a very useful and plausible, though incomplete account of cognitive processes that serve to justify moral decisions. He has managed to touch upon aspects of all three cognitive processes that characterize an ideal

---

[154] Ibid., 113.
[155] Ibid., 118.
[156] Ibid.
[157] Ibid., 127.

moral judge. What he failed to do was to develop them fully and explicitly as cognitive processes that would guide fair-minded individuals seeking to justify the morality of the decisions they are making.

Despite the merits of the procedures Daniels has put forward for making policy decisions, he makes assumptions and claims in his discussion of these procedures that definitely should be challenged, and in some instances, rejected.

Daniels does not attend to or enumerate the moral imperatives of justice that policy decision-makers should draw upon. One exception is his reference to the formal requirements of justice that similar cases be treated similarly. This formal requirement of justice is violated "if a coverage or risk protection decision disadvantages one group of people over others who are similar in all relevant ways…"[158] However, Daniels goes on to assert that being disadvantaged relative to others is a necessary outcome of decisions about which medical services should or should not be covered. Daniels only rejects disadvantages that are unnecessary since alternatives to them are available, as we have previously noted above. How then shall we understand his acceptance of inevitable inequalities in medical services as being justifiably condoned by organizations and agencies making coverage decisions.

First of all, he is making the assumption that, if resources are to be justly distributed in any given society, there are not enough resources available to assure that all medically indicated needs can be met. Now there are circumstances that create scarcities. For example, an unavoidably scarcity of physicians exists when a physician stops at the side of a road to lend aid to a number of victims of an automobile accident. He or she has no choice but to attend to the nearest one and so on sequentially. Only then can the physician depart from this customary lottery, namely first-come, first-served, and select on the basis of need anyone or more of the most severely injured for further assistance. Daniels acknowledges this resort to a lottery as a relevant one for patient selection, one that necessarily puts some patients in a position of being disadvantaged but which creates dissimilarities among them so that the situation created does not violate the formal principle of equality if justice is to be done.

What Daniels is overlooking, though, is that patients should be considered as strictly similar whenever their medical conditions and the need for treating them are similar. In other words, when it comes to coverage of individual medically indicated and available treatments there should be no decisions to exclude such coverage from anyone for financial reasons. If governments are to treat those under their jurisdiction justly, they will seek to find ways to assure that physicians can practice medicine in accord with the Hippocratic ethic. That means that those who cannot afford treatment, whether through lack of money for insurance, or for sufficient coverage from insurance companies or governmentally controlled programs, will receive the same treatment that all others with similar conditions receive. Daniels simply rules that out as viable and justifiable course of action.

Our major quarrel, then, with Daniels, has to do with his unquestioning and undefended conviction that no government or other entities have the resources to

---

[158] Ibid.

3.3 Inadequate Accounts of Moral Cognition: Sen, Nussbaum, and Daniels

provide everyone the medical care that is medically indicated and available, services that those with sufficient financial means can and do obtain. Daniels could express astonishment that anyone would be so uninformed as to be unaware of the unsustainable, ever increasing costs of health care. That unsustainability should be evident. Given present medical practices, Daniels is correct about that as we will document in subsequent chapters. But the question he should be asking is whether there are ways other than rationing care to reduce these costs, ways that are even morally imperative if justice is to be served. This is a question he did not ask. Daniels did not investigate and morally assess the extent to which enormously costly medical interventions presently in use are scientifically and morally justifiable.

Surely a treatise on achieving justice in the provision of health care should at least wish to ascertain the feasibility of reducing costs without curtailing necessary and effective medical care. The chapters that follow take on that task.

Chapter 4
Overdiagnosing, Overtesting, and Overmedicalizing Physical Conditions

Abstract This chapter documents two major contributors to the path of unsustainability within the U.S. healthcare system. First, is the sheer ever-higher-rising costs of medical services, and the relativity poor outcomes as compared with the world's other nations, all of whom spend considerably less. This is unsustainable. Second, the high costs that unethically contribute to unsustainability are: over diagnosis and the treatments that go with it (one who is over diagnosed can only be harmed, not benefited), and phenomena supporting over diagnosis are grounded in false notions of how to prevent diseases and conflicts of interest motivating medical experts and the FDA to keep adding to the numbers of individuals who allegedly qualify for being over diagnosed, and so over tested and over medicated

4.1 An Ethical Overview

In Chaps. 1 and 2, we identified some of the fundamental moral imperatives of justice. We say "some" because we are not claiming to have provided an exhaustive list of moral responsibilities and rights that may properly qualify as universally existing moral requisites of individual and communal life. By having recourse to justice in its generic sense, that is, as what we owe one another as human beings, we could uncover and defend the Hippocratic ethic, and the moral imperatives of justice that guided those whose practices were in accord with its admonishments.

To implement the Hippocratic notion of justice is to engage in meeting everyone's medical needs with appropriate, available care. Justice demands nothing less. Denying care to some that others with similar needs can and do receive is morally unacceptable: that moral judgment is to be found in Hippocratic treatises and medical practice.

Yet, as noted in Chap. 3, three prominent scholars, Daniels, Nussbaum, and Sen, all advocated that everyone be able to receive adequate medical care, something they described as a basic minimum. This they did in the name of justice. None of them even discussed the Hippocratic idea and practice of justice. We asked why they failed even to consider the Hippocratic ethic and its possible relevance. After all, this ethic has guided physicians for centuries and to some extent still does. In fact, as we shall be documenting in this chapter, American physicians are organizing themselves to increase compliance with one of the critical tenets of the Hippocratic ethic: protecting patients from harm.

As we discovered, these three scholars, like so many others who address health care policy, did not work with a generic concept of justice. That led them to overlook the very basic reasons for providing medical care, namely the moral impulses that move us to help and not harm those in need of assistance who are suffering, ill, disabled, or at risk of dying. They did not connect justice and empathy. Our empathic emotions express themselves in acts of compassion; they fuel our proclivities to nurture one another and rescue one another from suffering and death. Empathy makes it difficult to ignore anyone's need for care; it inhibits us from causing harm by inflicting it, or by failing to prevent or stop it. The Hippocratic ethic expects physicians to be empathic, be humane, toward those who are in their care, and all those who seek to be in their care.

But is universal health care financially sustainable? Sen does not directly address this question. Nussbaum thinks there is not sufficient support for anything more than a guaranteed basic minimum of care. Daniels alleges that meeting everyone's medical needs is not financially sustainable. For him, rationing medical care is financially necessary and just. For financial reasons, some medical care will have to be denied: a fair distribution of goods demands it. Adequate medical care for all has to be specified with this constraint in mind. The present unsustainability of current expenditures for medical care lends great plausibility to what Daniels is arguing for as a matter of policy. However, what Daniels, and for that matter Sen and Nussbaum as well, do not take into account, is that when the Hippocratic ethic called for denying care to no one, it specified the moral imperatives to follow that made this a financially feasible practice. As we shall observe, those moral imperatives will still help pay for medical care.

4.2 Why the American Health Care System Must and Ought to be Changed

To move away from the Hippocratic standards of what constitute ethically justifiable ways to provide medical care is to perpetuate the current perilous path to insolvency being taken by the medical care system in the United States. Michael Chernew and his co-authors, reflecting on the fact that health care spending historically exceeds the rate of income growth by more than 2 percentage points, conclude that, if this situation continues "the consequences for beneficiaries, federal and state budgets, and the entire economy—given the implied increase in tax rates and foregone

4.2 Why the American Health Care System Must and Ought to be Changed

consumption—will be dire."[1] As compared to other wealthy nations, the United States increased per capita health spending from 1970 to 2002

> At nearly twice the rate in other wealthy nations... As a result, the United States now spends well over twice the median expenditures of industrialized nations on health care and far more than any other country as a percentage of its gross domestic product (GDP).[2]

One might argue that spending more is beneficial. Indeed, Chernew and his co-authors, despite their alarm over the rate at which health care spending keeps growing assert that:

> Growth in health care spending is often a good thing. We want to spend more of our growing income on medical advances whose benefit exceeds their cost.[3]

The claim that growth in health care spending is often a good thing is not as self-evident as it appears to be for theses authors. To begin with, what qualifies as an advance medically is not always easy to ascertain. Be that as it may, let us consider an innovation that is generally agreed to be an advance. Antibiotics, for example, can save people from dying of life-threatening infections and diseases like pneumonia, far better than was possible before their advent: they can also clear up infections much more quickly than was previously possible. Antibiotics confer great benefits at a relatively low cost but they must be used properly in accord with our scientific knowledge of their usefulness. That's the rub! As we will later note, they have been excessively and unjustifiably resorted to at considerable cost, without benefit, and with some harm. Another consideration is that a medical advance, such as a drug, may have a favorable cost/benefit ratio but not one that is nearly as favorable for example, as eating nutritious food, getting sufficient rest, and taking time to exercise. As we will be discussing later, there are costly drugs that can be worth the cost if used appropriately and sparingly but whose excessive and scientifically unjustifiable uses not only drive up financial costs, but also cause harmful side effects. So whether growth in health care spending is often a good thing depends upon whether spending is carried out justly, that is, in the most beneficial, least costly, and least harmful way.

If growth in health care spending is _often_ a good thing, the United States should be obtaining the very best results in the world from the medical care it provides. Peter Muenning and Sherry Glied asked whether this is true: they looked at survival rates to assess this. What they found may come a surprise to many readers.

Although the United States spends more than twice as much per person as other industrialized nations, the inhabitants of these other nations enjoy 2.5 additional years of healthy life as compared with those residing in America. For example, Japan spent $2000 per capita in 2001 while the United States spent $5000, and yet

---

[1] Michael E. Chernew, Lindsay Sabik, Amitabh Chandra and Joseph P. Newhouse, "Ensuring the Fiscal Sustainability of Health Care Reform," New England Journal of Medicine 362:1, January 7, 2010, 1–3.

[2] Peter A. Muening and Sherry A. Glied, "What Changes in Survival Rates Tell Us About US Health Care," Health Affairs 29:11, November 2010, 2105.

[3] Chernew et alia, "Ensuring the Fiscal Sustainability of Health Care Reform," 1.

people in Japan attained 5.7 more years of good health.[4] Furthermore, with respect to life expectancy, the United States is falling behind other countries:

> In 1950, the United States was fifth among the leading industrialized nations with respect to female life expectancy at birth, surpassed only by Sweden, Norway, Australia and the Netherlands. The last available measure of female life expectancy had the United States ranked at forty-sixth in the world. As of September 23, 2010, the United States ranked forty-ninth for male and female life expectancy combined.[5]

Muenning and Glied found that there are sharply different responses to the fact that "Americans are spending more on health but living relatively shorter, less healthy lives."[6] Some see that as "evidence that the U.S. health care system is performing poorly."[7] Others attribute that to "high rates of smoking, obesity, traffic fatalities and homicides."[8] To discover which of these views is correct, or closer to the truth, Muenning and Glied focused on the period from 1975 to 2005 and examined two measures of health care system performance: cost and 15-year survival among men and women ages 45 and 65 in the U.S. as compared to 12 other industrialized nations.[9]

When the U.S. is compared to 12 other industrialized nations with respect to the mean gain made in 15-year survival rates during the period 1975–2005, this is what the study revealed: Forty-five-year-men ranked 8th and sixty-five-year-old men ranked 11th; forty-five year old women ranked 9th and sixty-five-year-old women ranked 13th.[10] During the 1975–1985 period, forty-five-year-old men had ranked 3rd and sixty-five-year-old men had ranked 8th; forty-five-year-old women had ranked 8th and sixty-five-year-old women had ranked 13th, a ranking that could go no lower and that never rose any higher.[11]

All U.S. declines in 15-year survival rates relative to 12 other comparable nations took place as U.S. costs per capita for providing health care "increased from rough parity with the highest spenders in 1975 to 50% higher than those of the next most costly system (Switzerland) in 2005.[12]

So the data show that the United States was decreasing its 15-year survival rates relative to 12 comparable industrial countries while greatly increasing its expenditures on health care much more than any other of these same countries. This leads some to posit the necessity of reforming the health care system in the U.S. But others counter that these relatively lower survival rates are largely due to high rates of smoking, obesity, traffic fatalities, and homicides all of which clearly adversely

---

[4] Muenning and Glied, "What Changes in Survival Rates Tell Us About US Health Care," 2105.
[5] Ibid.
[6] Ibid.
[7] Ibid.
[8] Ibid.
[9] Ibid, Exhibit 2, 2108.
[10] Ibid.
[11] Ibid.
[12] Ibid, 2107.

4.2 Why the American Health Care System Must and Ought to be Changed

affect survival rates. Muenning and Glied examined how the U.S. compares with other countries in these risk factors that decrease survival rates. Using cross-national data spanning three decades, they found that the "risk profiles of Americans generally improved relative to those of many other nations."[13] Yet in spite of this, and in spite of spending nearly twice as much on health care as other wealthy nations, America's relative 15-year survival rates declined. Hence it comes as no surprise to have Muenning and Glied concluding that: "the findings undercut critics who might argue that the U.S. health care system is not in need of major changes."[14]

At the conclusion of their article, Muennig and Glied suggest two possible changes to the existing U.S. health care system. They note that "unregulated fee-for-service reimbursement and an emphasis on specialty care may contribute to high U.S. health spending, while leading to unneeded procedures and fragmentation of care."[15] In their view, the reliance on these practices within the U.S. health care system may account for "both the increased spending and the relative deterioration of survival" they have observed.[16] And they claim, that if that is true, they are willing to bet that "meaningful reform will not only save money in the long term, it may also save lives."[17]

These authors have certainly identified one of the very serious problems plaguing the U.S. health care system, namely the rather extensive resort to unneeded medical interventions. What kinds of regulations of fees would at all be feasible and helpful are not matters they clarify. Clearly the extensive use of specialists' services does contribute to costly and unnecessary treatments, something we shall later document. However, the focus simply on fees-for-services and the services of specialists is far too narrow: it misses a whole host of changes that must and ought to occur in the U.S. health care system for the sake of financial sustainability and extending healthy lives. The focus should be on achieving justice as an overarching guide to what medical care people ought and ought not to receive. With the focus on achieving justice, a whole range of actors in the U.S. health care system are seen to be responsible for the extensive recourse to interventions that are highly expensive, unnecessary, and more harmful than beneficial or harmful only. The sources of such injustices include not only some of the choices being made by some physicians, but also some being made by the Food and Drug Administration and pharmaceutical companies. Both of these organizations contribute markedly to the expense and harmfulness of some medical practices: both also, in ways we shall document in this chapter and in Chaps. 5 and 6, all too often undermine adherence to treatments based on the best available scientific evidence. We are speaking of injustices in these instances because so much of the excessive and unjustifiable costs of current medical interventions result from practices and policies that violate a key requirement of being just: first of all do no harm.

---

[13] Ibid, 2105.
[14] Ibid.
[15] Ibid, 2112.
[16] Ibid.
[17] Ibid.

American medicine presents us with a momentous paradox: never before have American physicians so abundantly possessed such exquisite skills and resources to save lives, and restore and enhance the health of those they serve so conscientiously. Yet, at the same time, the very power of contemporary medicine to save lives, prevent diseases, and restore, assist, and supplement vital human functions, is also a power which, if overused or otherwise misused, can debilitate, maim, sicken and even destroy lives. Paradoxically then, the injunction to first of all do no harm is an ethical imperative that has come to be too often violated by the overuse and misuse of the highly refined capabilities of contemporary medicine, capabilities for which we can all only be grateful whenever we are recipients of their appropriate use.

4.3 Organized Efforts by American Physicians to Increase the Just Delivery of Medical Care

Since American physicians are generally conscientious and are intent upon the pursuit of the well being of their patients, no one should be surprised that American physicians are now making a concerted effort to confront the very serious situations being created by the overuse and other misuses of quite a number of medical procedures. To begin with, there is a growing movement of medical societies that has organized itself under the banner of "Choosing Wisely."[18] Nine medical specialty societies initially banded together and the movement keeps growing as ever more specialty societies have joined.[19] The initial focus is precisely upon the overuse and other misuses of medical resources which the organizers rightly perceive as "a leading factor in the high level of spending on health care" and as a set of practices that submit patients to being at risk of harm.[20]

The Choosing Wisely campaign is a very significant and promising development, especially so because physicians account for about 80% of health care expenditures, and because some estimates suggest that 30% of all health care spending is wasted.[21] As Dr. Jay Siwek, the editor of American Family Physician notes in his endorsement of the Choosing Wisely initiatives, physicians "often order tests or treatments that do not stand up to clinical scrutiny."[22] So now physicians who have benefited from practice guidelines concerned mainly with what to do need to gain more knowledge of what not to do, that is knowledge of what tests and treatments

---

[18] Christine K. Cassel and James A. Guest, "Choosing Wisely: Helping Physicians and Patients Make Smart Decisions About Their Care," J.A.M.A. 307:12, May 2, 2012, 1801–1802.

[19] The names of the medical societies that have joined the Choosing Wisely movement and the lists of changes in medical practices they are suggesting are available at www.choosingwisely.org

[20] Cassel and Guest, "Choosing Wisely," 1801.

[21] Ibid.

[22] Jay Siwek, "Choosing Wisely: Top Interventions to Improve Health and Reduce Harm, While Lowering Costs," American Family Physician, 86:2, July 15, 2012, 128.

not to offer because they do not help their patients and can cause them harm.[23] Siwek has presented [two] tables listing a number of medical interventions that should be curbed or dropped altogether when appropriate.[24] To illustrate how these curbs on unnecessary and harmful treatments on this list can reduce costs, consider a simple condition like sinusitis. Though it usually resolves itself untreated, antibiotics are being wrongly prescribed in more than 80% of outpatient visits for acute, moderate sinusitis. This costs a lot of money: sinusitis accounts for 16 billion office visits and 5.8 billion dollars in expenditures annually. Siwek provides no estimates for them but one can discern that the other unnecessary and misguided procedures on his list are very costly as well.

The Choosing Wisely campaign is clearly on a path to increase the extent to which American physicians are adopting guidelines for medical decisions that are in conformity with the moral imperatives of justice. The medical decisions being advocated by the various medical specialty societies are all designed to reduce tests and interventions that are often unnecessary. (To obtain an up to date list of these guidelines visit www.choosingwisely.org) Interventions being curbed or eliminated are those that cannot be justified clinically: they are of no evident benefit and can only harm. What is being curbed or eliminated, as the case may be, will improve health care quality, reduce harm to patients, and, at the same time, greatly lower costs. Reducing costs is not only beneficial for patients as such but also for insurers and taxpayers. Given that these cost reductions result from better medical care rather than the denial of needed care, they are morally justifiable: they are just. Furthermore, these initiatives of the Choosing Wisely endeavor put considerable and definite emphasis on avoiding harms, particularly those associated with exposure to radiation. In this respect, Choosing Wisely is also a matter of choosing justly. First of all do no harm is a time-honored maxim of medicine; it is one of the hallmarks of the Hippocratic conception of justice.

Among the noteworthy ways in which unnecessary interventions are being identified for the purpose of avoiding them, is that of having physicians determine if patients are free of symptoms and devoid of risks for developing symptoms, and thus they are not candidates for being tested and/or treated. This begins to address a very serious problem. The American College of Cardiology, for example, is advising physicians not to perform "stress cardiac imaging or advanced non-invasive imaging in the initial evaluation of patients without cardiac symptoms unless high-risk markers are present." Why? "Asymptomatic, low-risk patients account for up to 45 percent of unnecessary 'screening.'"[25] This points to a pervasive problem plaguing American medicine today.

---

[23] Ibid.
[24] Ibid, 130 and 133 […] in Appendix II.
[25] Visit www.choosingwisely.org

4.4 Overdiagnosing and Overmedicating: Practices that Need to Be Curbed

Diagnosing and treating individuals who are free of symptoms and not at risk for developing them has become a major source of needlessly harming patients and driving up costs. H. Gilbert Welch, a physician and professor at the Dartmouth Institute for Health Policy and Clinical Practice, refers to this phenomenon as "overdiagnosis" and describes it as follows: "overdiagnosis occurs when individuals are diagnosed with conditions that will never cause symptoms or death."[26] Welch notes that "the conventional wisdom is that more diagnosis—particularly more early diagnosis—means better medical care."[27] The logic of these practices is that "more diagnosis means more treatment and more treatment means better health."[28] This may be true sometimes but, as Welch tells us,

> Excessive diagnosis can literally make you feel sick. And more diagnosis leads to excessive treatment—treatment for problems that either aren't that bothersome or aren't bothersome at all. Excessive treatment, of course, can really hurt you. Excessive diagnosis may lead to treatment that is worse than the disease.[29]

There is this enthusiasm, all too rampant, of seeking early diagnosis either in connection with organized screening efforts or routine exams, and that results in considerable overdiagnoses.[30]

As we have observed and illustrated, the Choosing Wisely lists sometimes reject and sometimes reduce some current instances of overdiagnosing. On the whole, the Choosing Wisely lists advocate clinical decisions that are on very sound footing. To the extent that they are implemented, they will definitely improve the quality of medical care and reduce medical expenditures. Nevertheless, there is more room for more efforts to curb overdiagnosing and overmedicating patients. Indeed, our discussion of some such efforts needed will begin with two approaches to medical decisions that appear on the lists submitted by Siwek. Hopefully, the medical societies taking the approaches in question, will welcome the suggestions supported by scientific studies we will be presenting.

---

[26] H. Gilbert Welch, Lisa M. Schwartz, and Steven Woloshin, Overdiagnosed: Making People Sick In the Pursuit of Health (Boston, Massachusetts: Beacon Press, 2011), xiv.
[27] Ibid.
[28] Ibid.
[29] Ibid.
[30] Ibid.

4.4.1 Osteoporosis: A Case of Misguided Practices

Osteoporosis refers to a condition of having bones that are regarded as abnormally thin, that is to say, thin enough to pose an unacceptable risk of falls that cause bone fractures. Falls that result in hip fractures are particularly worrisome: they can be lethal, especially in older individuals. In their efforts to eliminate unnecessary procedures, the societies of family and internal medicine both recommend some limits to screening and treating individuals for osteoporosis. To be excluded from screening, are women who are under the age of 65 and men under the age of 70 who are free of certain risk factors.[31] However, they do approve screening women 65 and older and men 70 and older. Their stated rationale for doing so is that it is cost-effective, whereas screening younger low-risk individuals is not.[32] This rationale is what we will now carefully question. In the light of the available scientific studies, what should be questioned is whether screening anyone for osteoporosis is cost-effective and harmless enough to justify, given the existence of more effective, less costly interventions that are free of side effects and that prevent hip fractures, and fractures more generally.

To begin with, the very definition of osteoporosis is problematic. An x-ray is taken of bones to measure their density. A so-called "T score" is used to quantify the results. A normal score of "0" is attained when it corresponds to the average bone density of the typical white woman 20–29 years old and that regardless of the age or ethnicity of the individual being tested. If the tested person's bones are a great deal denser than average the T score could be as high as 3; if a person's bones are a whole lot thinner the T score could be as low as −3. Bones thin as one ages so older women typically have T scores below 0. When, then, should an individual be treated? The World Health Organization (WHO) set the bar at a T score of less than −2.5. Viewing this definition as arbitrary, and assuming the increased risk of fractures the more the T scores fall below 0, the National Osteoporosis Foundation, in 2003, advocated treating all women who had T scores of less than −2.0. Immediately, the number of women labeled as having osteoporosis went from 8,010,000 to 14,791,000 an increase of 6.7 million American women deemed to require treatment.[33]

How effective is it to increase one's bone density for the purpose of preventing falls? Welch has provided us with an analysis of the "Fracture Intervention Trial," a randomized study of women, typically 68 years old, designed to ascertain "the effect of increasing near normal bone density in women who had not had fractures previously."[34] Based on the data from this study, Welch calculated what would happen if 100 of these women were treated for a lifetime: five of them would be helped because they would avoid fractures; forty-four of them would be treated in vain

---

[31] Siwek, "Choosing Wisely," 130.
[32] Ibid.
[33] Welch, Overdiagnosed, 22–23.
[34] Ibid, 26.

because they experience fractures despite receiving treatment and also may experience side effects from being treated; fifty-one lose out completely, having been overdiagnosed.[35] Imagine 95% of patients, being treated for a lifetime once they are in their sixties, incurring the expense of treatments and being subject to the risks of serious side effects, while receiving no help whatsoever! The drugs now most commonly used are biophosphonates. They can make bones more brittle, disturb calcium metabolism, lead to ulcers in the esophagus, and very rarely, cause bone to die. Indeed, whether bone density can be increased too much is actually not even known. Despite these side effects, people are receiving messages that overstate the benefits of being treated and largely have nothing to say about the potential harms. This leads to an exaggerated view of reality, namely that it is safer to be diagnosed and treated than to live with the fear of an increased risk of fractures associated with aging.[36]

Welch is concerned about yet another change in the guidelines for treating osteoporosis promulgated by the National Osteoporosis Foundation. The T score cut off for treatment has been expanded to -1.0 but the T score alone is not by itself a warrant to treat someone. Rather the physician is to enter a patient's age, weight, height, and T score into a WHO website and find definitions of the following risk factors for experiencing a fracture: smoking, using steroid medications, a prior history of fracture, disorders associated with osteoporosis, and having three or more alcoholic drinks per day. The physician then gleans any such data through an interview and provides it to the WHO computer program designed to calculate whether the patient in question has a higher than 3% chance of having a hip fracture in the next 10 years. If so, the patient should be treated.[37]

While Welch commends efforts to better define who is at risk for fractures, the recommended procedures are, in his view, problematic. First of all, we do not have any knowledge as to whether this refinement helps because we lack any evaluation of treating women with other risk factors who have a close to normal bone density as indicated by a T score of -1.0.[38] Secondly, the recommendations are so complex and time consuming that he does not expect many physicians to follow them: they will simply treat every woman with a T score below -1.0. That would mean that almost all older women will be regarded as in need of treatment. Applying this set of procedures to men raises the same troubling questions.[39]

As already noted earlier, the specialty groups, Family Medicine and Pediatrics, consider it to be cost-effective to screen women 65 years of age and above and men 70 years of age and above. That is a rationale they adopt, one endorsed by the American Association of Clinical Endocrinologists, the American College of Preventive Medicine, the National Osteoporosis Foundation, and the U.S. Preventive

---

[35] Ibid.
[36] Ibid, 27.
[37] Ibid, 28–29.
[38] Ibid, 29.
[39] Ibid.

4.4 Overdiagnosing and Overmedicating: Practices that Need to Be Curbed

Services Task Force.[40] Family Medicine and Pediatrics also advocates screening younger women who are 50–65 years old, but only those with certain conditions that they regard as putting these women at risk for fractures. These conditions "include, but are not limited to, fractures after 50 years of age, prolonged exposure to cortisteroids, diet deficient in calcium and vitamin D, cigarette smoking, alcoholism, and thin—small build."[41] These risk factors are partly different from those that appear in the treatment recommendations made by the National Osteoporosis Foundation as cited above. Hence, we cannot tell how much physicians in specialties of Family Medicine and Pediatrics are following those guidelines for determining when to treat, such as the T score of −1.0 as the cut off for considering whether to treat those with certain risk factors.

Be that as it may, the research as analyzed by Welch confronts individuals with a very difficult choice as to whether to accept treatment for osteoporosis, since 95% do not benefit and my suffer serious side effects, and the side effects one risks may appear harmful enough to outweigh the benefit treatment offers. Fortunately, given the seriousness of suffering fractures, particularly hip fractures, there are scientific studies of interventions that provide sound reasons for abandoning the present policies of screening and using drugs to treat osteoporosis: as these studies reveal, fractures can be prevented by interventions that are much more beneficial, much more effective, and much less costly than those currently in vogue, and with the added bonus that they are free of harmful side effects.

In 2004, John Abramson, identified as a physician on the clinical faculty of Harvard Medical School, published Overdosed America, a work aimed at overcoming the overuse, costliness, and harmfulness of medications and the tendency to overlook safer, less costly and more effective modes of preventing and treating illnesses.[42] The practices associated with osteoporosis are among those that Abramson chose to address.

Having discussed the same research analyzed by Welch, Abramson takes note of two other studies. One published in 2001 showed that women between the ages of 70 and 79 with severe osteoporosis yielding T scores of −4 or − 3, and with a major risk factor for hip fracture, benefited little from being treated with the drug Actonel. Those with a pre-existing spinal fracture, 40% of those in the study, did experience a slight reduction in hip fractures. It was slight in the sense that one hundred women would have to take Actonel for 1 year to prevent one hip fracture. The other 60% of the women studied experienced no significant reduction in their risk for hip fractures. Noteworthy also is that those who took the drug and those who took the placebo did not differ at all with respect to their overall health as measured by the number of serious illnesses (causing hospitalization or death) including fractures that they experienced during the three-year study.[43]

---

[40] Siwek, "Choosing Wisely," 130.

[41] Ibid.

[42] John Abramson, Overdosed America: Broken Promise of American Medicine (New York: Harper Collins, 2004).

[43] Ibid, 214.

In the second study published in 1999, the very same results were found in younger women who had an average age of 69, had suffered at least one spinal fracture and had been diagnosed with osteoporosis. Those who took the drug Actonel had fewer fractures but no reduction in the occurrences of serious illnesses as compared with those who did not receive medication. Abramson concludes his assessment of these two studies with this strongly critical comment: "The net effect of drug treatment on the risk of serious illness in the highest risk women? Nothing—except the cost of the drug."[44]

The ineffectiveness of drugs being used to prevent hip fractures was even more startling for the women over the age of 80. There were 3880 of them that participated in the 2001 study discussed above. Treatment of these women with Actonel had "no effect on the incidence of hip fracture."[45] Of these women, 80% had been diagnosed with osteoporosis and the rest had at least one risk factor for falls. These results are rather tragic considering that two of three hip fractures occur in women who have reached the age of 80. Since 90% of hip fractures occur as a result of falls, one would expect that the oldest and frailest women would be at the greatest risk.

The rather meager help that is being provided by bone mineral density (BMD) testing, and subsequent treatment with drugs should have been anticipated or at least suspected. On what basis? There was a study carried out in the Netherlands, reported in 1997, that documents the limitations of BMD testing.[46] What the study found was that BMD testing identified no more than one-sixth of the risk of hip fractures for women between 60 and 80 years of age. Just as important as a T score were other factors: increased frailty, muscle weakness, side effects from drugs, declining vision, and smoking cigarettes. For Abramson, this research questions what the WHO's definition of osteoporosis has initiated, namely a mistaken focus by physicians and patients on "the results of BMD testing as the sole or primary predictor of fracture risk." And as he goes on to remark, "routine BMD testing may not be the best way to help women prevent hip fractures, but it is an excellent way to sell more drugs."[47]

Abramson very definitely views present policies of screening for osteoporosis, followed by prescribing drugs for those deemed to have osteoporosis, as seriously flawed indeed misguided. Yet he also regards fractures, especially hip fractures as very deleterious threats to wellbeing and to life itself. For that reason, he cites numerous studies of practices that do prevent falls and fractures far more effectively, for much less cost, without harmful side effects while at the same time, contributing to improving overall good health.

To build and maintain strong bones, proper exercise and good nutrition are essential throughout various stages of life. There is specific scientific evidence cited by Abramson that exercise builds up the very bone structures that support the vulnerable

---

[44] Ibid.
[45] Ibid, 215.
[46] Ibid, 214.
[47] Ibid, 215.

areas of the skeleton in later life.[48] With respect to preventing hip fractures, the NIH sponsored research, the Study of Osteoporosis Fractures, a study of almost 10,000 women aged 65 and older who were living independently.[49] The results indicated that, over a 7-year period, those women who exercised moderately had a statistically significant 36% fewer hip fractures than those who exercised least. This reduction in absolute terms was twice that achieved by using the drug Fosamax in the study published in JAMA in 1998, the one cited and analyzed by Welch for its mineral benefits and harmful side effects and discussed earlier above.

In addition, there is a randomized study conducted in Sweden, published in 2002, that tested the efficacy of a program to prevent falls.[50] The subjects were nursing home residents, 83 years old on average. The program included exercise, medication review, hip protectors, and staff conferences on reducing the risk of any repeat fall. During the 8-month program, only 1.6% of those in the fall-prevention program experienced hip fracture while 6.1% of those in the control group had a fracture. Abramson understandably calls this a "dramatic reduction" and reminds his readers that "Actonel did not reduce hip fractures in women of similar age."[51]

Abramson briefly refers to several more studies that reveal ways to reduce the risk of falls and hip fractures. One is to increase a person's strength. Tai Chi, a form of exercise, improves one's balance and it has been shown to cut the risk of falls in half for people who are 70 years old and older. Diet is also significant. To build and sustain strong bones, a daily adequate intake of calcium, 1200 mgs., usually no more than 1000 mgs. From supplements would be required, and 400 to 800 IUs of vitamin D. It has also been shown that diets with a higher ratio of animal to vegetable protein raise the rate of bone loss in women 65 and older. Indeed, it has been observed that "women whose diets contained the highest proportion of animal protein were almost four times more likely to suffer a fractured hip than women whose primary source of protein was vegetables."[52]

For Abramson this is but a sampling of some of the research the public will not find out about from commercial sources of information about how to attain and maintain healthy bones. For example, at the website that Merck, the manufacturer of Fosamax sponsors you will be told to obtain your T score and, if it falls below −1.0, find out about treatment options from your doctor.[53] The information you would obtain from the National Osteoporosis Society will also not be devoid of commercial influence: it is a tax-exempt, nonprofit organization and the recipient of a large amount of financial support from drug companies.[54]

There is actually research published in 2004 in the International Journal of Technology Assessment in Health Care that reveals how difficult it is to secure

---

[48] Ibid, 217.
[49] Ibid, 217–218.
[50] Ibid, 218.
[51] Ibid.
[52] Ibid.
[53] Ibid.
[54] Ibid, 219.

unbiased information from the Internet.[55] The study compared the information on bone mineral density testing found in the most frequently visited consumer health websites with information available on the websites of health technology assessment websites that are not commercially funded. On the primarily commercially sponsored consumer health sites, the consistent message conveyed was that BMD testing simply and painlessly predicts one's risk of fracture from osteoporosis. However, the message conveyed on the websites of health technology assessment organizations was very different but no less consistent: BMD measurements are not good predictors of fracture risk."[56]

On the basis of the research he has reviewed, Abramson concludes that the drug companies have been successfully "disease mongering" by planting the fear in women (and now men also), who had thought themselves to be healthy. The fear that they are in danger of having their bones break suddenly and without warning. Writing before men were being targeted as well as women, Abramson has this remedy to offer for preventing falls and fractures: "All post menopausal women should be exercising routinely, eating a healthy diet, taking calcium and vitamin D supplements, and decreasing their risk from falls."[57] To make these recommendations does not compel anyone to offer or receive screening for osteoporosis. Consider how much these preventive measures, if followed, have to commend them: they quite effectively prevent fractures; unlike drugs, they are free of any harmful side effects and cost a great deal less to carry out; and they increase one's overall health.

One would think that the physicians in the societies of family and internal medicine would find Abramson's approach to prevent falls and fractures congenial and worth implementing. After all, they are committed to practicing medicine in ways that employ the least costly, least harmful and most effective means to care for their patients. At the very least, these aims should lead them, before any resort to drugs, to seek to gain compliance from their patients: to avoid smoking and excessive consumption of alcohol; to exercise, to eat nutritious food—particularly fruit and vegetables; and, especially as they age, to reduce the risks of falling and be sure to obtain enough calcium, magnesium, and vitamin D to overcome any deficiencies in these nutrients, vital for maintaining strong bones and good health generally. Perhaps they do this. If so, one would hope they would also seriously consider the very harmful side effects of the relatively ineffective drugs presently being used to prevent fractures, and decide to greatly curtail their use, or even avoid resorting to them altogether.

---

[55] Ibid.
[56] Ibid.
[57] Ibid, 219–220.

4.4.2 Heart Disease: A Case of Excessive Resort to Statistics

In 1957, the Framingham Heart Study, begun in 1948, published its results: among its findings was that of a link between elevated levels of cholesterol and an increased risk of heart disease.[58] A description of the different effects of HDL and LDL cholesterol emerged in 1977. LDL cholesterol was called "bad" because of its role in building up plaque in one's arteries, and that is a process, if left to continue unabated, that can block the arteries and trigger a heart attack. HDL cholesterol, on the other hand, is called "good" because it acts to pull cholesterol out of the arteries and transport it back to the liver where cholesterol is manufactured. The theory developed that lowering cholesterol with medication will lower the risk of coronary heart disease by reducing plaque formation.[59]

The current mode of treatment using statins began in 1987 and it is this mode of treatment that is endorsed by the physicians in internal medicine. In their commitment to "Choosing Wisely," they propose that only generic statins be used to the extent that they are effective.[60] Doing this will help meet their goal of reducing the costs of medical care, considering that in 2011 there were more than 20 million Americans taking some form of statin: the statin Lipitor is the all-time biggest selling prescription medicine having achieved cumulative sales exceeding 130 billion dollars.[61] One can readily see the desirability of reducing costs by taking advantage of its cheaper generic versions of statin, now available since late in 2011.

However, there are physicians who are convinced that the scientific evidence does not justify the extensive use of statins to treat heart disease. Among these is Dr. Norton Hadler, professor of medicine and microbiology/immunology at the University of North Carolina, Chapel Hill, and attending rheumatologist, University of North Carolina Hospitals. He acknowledges that: cholesterol is a risk factor; statins can lower cholesterol, and that doing so, will slightly decrease the likelihood that someone who has had a heart attack will suffer another one and increase to a barely measurable degree their survival rate. Having granted that much, he is unconvinced that "statin treatment affords any meaningful advantage to people who have not had a heart attack."[62]

This view of the relative ineffectiveness of statins to prevent heart disease deserves our attention because there is research that reveals much less costly and more effective ways to accomplish this, and that do so without causing the serious side effects that result from taking statins. We turn now to consider studies that document these claims.

---

[58] Ibid, 131–132.
[59] Ibid, 132–133.
[60] Siwek, "Choosing Wisely," 130.
[61] Bill Berkrot and Ransdell Pierson, "FDA Adds Diabetic, Memory Loss Warning to Statins," Reuters in New Hampshire Union Leader. February 29, 2012, B7. In 2011 a generic version (atorvastatin) became available, costing far less.
[62] Hadler, The Last Well Person: How to Stay Well Despite the Health-Care System. (Montreal, Quebec: McGill-Queens University Press, 2004).

4.4.2.1 The Framingham Heart Study

Abramson asks why it is that cholesterol gets so much attention in the efforts to prevent heart disease and strokes.[63] He notes, at the outset of pursuing this question, that cholesterol is not in itself a health risk but rather a vital ingredient in maintaining many of the body's essential functions. A very significant example of this fact is that "cholesterol is the most common organic molecule in the brain," and as Abramson observes, "this could explain why statins have a small but statistically significant negative effect on cognitive function."[64] Furthermore, cholesterol is needed for the maintenance of a number of important hormones such as blood-sugar regulating hormones and sex hormones.[65] It is not surprising, therefore, that the side effects of statins include elevated blood sugar resulting in a diagnosis of type 2 diabetes and an increase by about 50% of the frequency of sexual dysfunction in men.[66]

Abramson reminds his readers that, in the light of the vital role of cholesterol for sustaining our bodily functions, the real goal of medical care is not simply to lower blood levels of LDL cholesterol, but to decrease the risk of coronary heart disease, all kinds of serious illnesses, and premature death. In short, the goal of medical care is to improve overall health.[67]

With that goal in mind, how well does lowering cholesterol do as viewed from a scientific perspective? Given what he refers to as "the current cholesterol-lowering craze," Abramson finds it alarming that the Framingham Heart Study showed that the risk of death <u>increases</u> significantly with lower total cholesterol levels for men and women after they reach the age of 50."[68] Citing additional data from the Framingham, Abramson notes that "physical activity, unlike total cholesterol levels, is highly correlated with overall mortality rate: the most active third of the original 5000 men and women in the study had a 40 percent lower death rate than the least active third."[69]

Despite the negative effects of lowering total cholesterol, doing so has become a very important preventative health care strategy. Indeed, the pool of those for whom lowering their cholesterol has been deemed to be advisable and necessary has been greatly expanded. In the mid-1990s large health-care organizations adopted a total cholesterol over 240 as the criterion for an abnormality that warranted therapeutic intervention: it had been pegged at over 300. Then, in 2001, a panel of experts published guidelines that set the threshold for abnormal cholesterol levels at greater

---

[63] Abramson, Overdiagnosed America, 134.
[64] Ibid.
[65] Ibid, 133–134.
[66] Berkrot and Pierson, "FDA Adds Diabetes, Memory Loss Warnings to Statins," B7. See also E. Bruckert, P. Giral, H.M. Heshmati, and G. Turpin, "Men Treated with Hypolipidaemic Drugs Complain More Frequently of Erectile Dysfunction," Journal of Clinical Pharmacology and Therapeutics, 21, 1996, 87–94.
[67] Abramson, Overdosed America, 134.
[68] Ibid.
[69] Ibid.

than 200. That is close to the average level in the U.S. adult population.[70] That increased the number of diseased Americans from 49,480,000 to 92,127,000.[71] With over 20 million individuals already taking statins in 2011,[72] and the potential for many more doing so, one does have ample reason to ask whether there is scientific evidence for justifying such a considerable resort to lowering cholesterol and for using statins for that purpose. We begin by considering two major studies that were cited to justify the recommended increase in 2001 of those who should receive statins.

4.4.2.2 The West of Scotland Coronary Prevention Study (WASCOPS)

This study was undertaken to see if statins could prevent heart attacks in a region with one of the highest rates of heart disease in the world.[73] The results obtained had the effect of increasing the consideration given to statins by public health policy and by recommendations put forward by advisory panels such as those of the American heart Association.[74] This was a 5-year randomized trial in which just over half of the men in it took 40 mg. of pravastatin, while the others took a placebo.[75] These men were at high risk of having a heart attack since their average LDL cholesterol level was high at 192 mg/dL; 44% of them smoked; and one out of five had symptoms of blocked arteries.[76]

Hadler does not find the results to be such that well people should feel compelled to be screened for their level of cholesterol to find out if they should be treated with a statin. First of all, high doses of pravastatin did not save any lives since, as he notes, "the difference between the numbers who died from any cause is neither statistically nor clinically meaningful."[77] Nor was there any difference in the likelihood of deaths from non-cardiovascular causes: Hadler considers that important because "death from strokes were not avoided."[78] Originally there was no statistically significant difference between the percentage of men on placebo and the percentage of men on pravastatin who suffered a fatal myocardial infarction. Subsequently, the investigators shifted ten deaths into the heart attack category on the supposition that these might have died from a heart attack. It was then that a barely statistically significant difference was achieved, 4.2 times out of hundred, 5

---

[70] Welch, Overdiagnosed, 20–21.

[71] Ibid, Table 2.1, 23.

[72] Berkrot and Pierson, "FDA Adds Diabetes, Memory Loss Warnings to Statins," Union Leader, February 29, 2012, B7.

[73] Abramson, Overdosed America, 136.

[74] Hadler, The Last Well Person, 35.

[75] Ibid. Hadler puts the number of subjects in the study at 6595; Abramson, Overdosed America, 136 puts the number at 6600.

[76] Abramson, Overdosed America, 136.

[77] Hadler, The Last Well Person, 36.

[78] Ibid.

times out of hundred being the agreed upon cutoff for statistical significance. On that basis, the authors of the study claimed that pravastatin saved lives.[79] Hadler, however, disputes this claim. Even if one grants the questionable shift of ten deaths to the category of deaths caused by a heart attack, a difference of 0.6% of deaths attributed to heart attacks between those who took a placebo and those who took pravastatin is not, in Hadler's view, clinically meaningful. There are attributes of the men who are being studied that are hidden. Some are yet undefined such as genetic factors; others such as an individual's coronary artery anatomy cannot justifiably be determined since that would entail exposing over 6000 well men to the hazards of cardiac catheterization. Disparities between the two groups on such variables could very well account for a 0.6% difference in outcome, a difference therefore too tiny to be clinically meaningful.[80]

Another way of looking at the benefits of taking pravastatin is to note that 100 men in this study would have to take the drug for 2 full years to prevent a single heart attack, and five and a half years to prevent a single death from a heart attack.[81] Since over the 5 year period of the study, there was no statistically significant difference in the number of deaths experienced by those taking the statins and those receiving a placebo, it is extremely important to consider the full range of side effects from taking statins. We will do so once we have finished assessing their alleged benefits.

4.4.2.3 The Air Force/Texas Coronary Atherosclerosis Prevention Study (AECAPs/TexCAPS)

This is the second study of using statins to prevent heart attacks and deaths from them cited by the expert panel in 2001 that reduced the threshold for abnormal levels of cholesterol from above 240 to 200. This study, unlike the West Scotland Study, involved subjects who had what was then considered slightly below the normal level of cholesterol, 228, and lowering that to 184 for those receiving statins.[82] These subjects constituted 6600 healthy middle-aged and older people who were treated with lovastatin or a placebo for 5 years. The two groups were compared with respect to the frequency of several health outcomes: "occurrence of coronary heart disease, any serious disease, death from coronary heart disease, and death from any cause."[83]

In the 2001 guidelines, the results of this study are summarized as showing that LDL-lowering therapy in individuals with borderline-high-LDL cholesterol levels results in a large reduction in relative risk. As Abramson points out, they do not say

---

[79] Ibid, 36–37.
[80] Ibid, 37–38.
[81] Abramson, Overdosed America, 136.
[82] Welch, Overdiagnosed, 21.
[83] Abramson, Overdosed America, 137.

what risk the statin reduced.[84] It is true that those who took the statin had a 37% significantly lower risk of developing heart disease than those who received a placebo. This looks like a large difference when considering the relative risk. However, Welch notes that over 5 years 5% of untreated subjects had suffered first acute major coronary events as compared with 3% of those on the statins. That means that, for every one hundred patients treated for more than 5 years, only two were helped in this regard.[85] One has to add "in this regard" because of what Abramson says about the overall effects of being treated with statins: the risk of developing any serious disease (the kind of illness that requires hospitalization or that causes death) was identical in the people who took the statins and those who took the placebo."[86] Those who set up the new guidelines of 2001 nevertheless reported that no conclusion can be drawn about the effect of statin treatment on overall risk of death. Abramson categorically denies this. In this study of 6600 individuals with somewhat elevated LDL cholesterol, being treated with a statin did not decrease overall mortality. As a matter of fact, 80 of the people receiving statins died, as compared with 77 who received placebos. "In other words, "Abramson asserts, "the net result of treating people with moderate risk of developing coronary artery disease with a statin was simply to trade coronary heart disease for other serious diseases, with no overall improvement in health."[87]

Abramson analyzed the additional studies the published guidelines relied upon to recommend statins for individuals who had cholesterol levels above 200. He found no data that supported treating healthy people with statins. He did find that statins decrease subsequent heart attacks. Abramson is inclined to treat men and women who have heart disease with statins.[88] This, of course, falls far short of the extensive statin usage recommended by the expert panel in its guidelines of 2001.

4.4.2.4 The ALLHAT Study

This study began in 1994 and was published in 2002.[89] Of the more than 10,000 patients consisting of an equal number of men and women age 55 and older, 90% of the men and 75% of the women qualified as having risk factors for statin therapy as stated in the new guidelines. The patients in the study were randomly assigned, either to receive pravastatin or to receive the care from their own doctors typically offered at that time. By the time the study ended, 83% of those assigned to take pravastatin were still doing so and 25% of those cared for by their doctors had been switched to statin therapy. Hence, the study tested what would happen if the number of Americans taking statins was tripled using the 2001 guidelines.

---

[84] Ibid.
[85] Welch, Overdiagnosed, 21.
[86] Abramson, Overdosed America, 137.
[87] Ibid, 138.
[88] Ibid, 138–143.
[89] Ibid, 144.

As it turned out, tripling the number of people on statins did not prevent heart disease and did not decrease overall risk of death. This increase in the number of patients given statins beyond the norm in place in the mid-1990s was of no benefit for people age 55 to 64 or 65 and older, for men or women, for those with or without diabetes, or those with LDL cholesterol higher or lower than 130 mg/dL. African Americans did have fewer episodes of heart disease but experienced no fewer deaths.[90]

These findings did not support eh 2001 guidelines. Yet in the medical journals the results of this study were largely rejected by experts who took the view that so many of the usual-care group had been put on statins that the difference in cholesterol levels between those on pravastatin and the usual-care group was not sufficient to reveal the benefit of statins. Abramson sees this as missing the whole point of the study. High-risk patients treated by their doctors using the standards of the mid-1990s were obtaining the maximum benefit statins have to offer. Tripling the number of individuals receiving statins, and doing so almost precisely in line with what the new guidelines recommend, yielded no further benefit.[91] In short, the ALLHAT study reveals the lack of scientific evidence for implementing the 2001 guidelines for the use of statins.

4.4.2.5 The Harmful Side Effects Caused by Statins

There are further reasons for questioning the 2001 guidelines and the currently extensive resort to statins. To begin with, consider the PROSPER study (Provastatin in Elderly Individuals at Risk of Vascular Disease).[92] This was a study seeking what the effect of stating therapy would be on high-risk elderly patients between the ages of 70 and 82. The results revealed that statins did not reduce the risk of developing heart disease or stroke for those who did not have heart disease to begin with: however, their risk of developing cancer increased significantly ($p > .02$). Moreover, this was a risk that increased each year the subjects took statins. By the fourth year of the study there resulted more than one extra case of cancer for every 100 individuals receiving statins each year.

The 2001 cholesterol guidelines categorically denied that there was evidence suggesting that cancer as a side effect of taking statins was at all a possibility. For whatever reason, the panel of experts overlooked or ignored the research published in 1996 that found that statins cause cancer in rats.[93] Cancer in humans can take some time to develop so there should at least be concern about cancer as a possible side effect for those taking statins for longer periods than the average 5 year stretches typical of existing studies of statin therapy.

---

[90] Ibid.
[91] Ibid, 144–145.
[92] Ibid, 145.
[93] Ibid.

Another very serious side effect, highlighted by Hadler, is that statins can cause a serious destruction of muscles that sometimes proves to be fatal: fifty or more such fatalities forced one statin, Baycol, from the market. Other statins have reported fewer such catastrophic side effects. Yet they do occur from being exposed to each of the statins along with cases of muscle toxicity that is milder and reversible. Hadler considers the benefits of statins too slight to tolerate much risk.[94] Citing the results of the three studies reviewed above, Hadler concludes that, when it comes to using statins to prevent heart disease and strokes in well people, a large number of them given the 2001 guidelines, "statins should be viewed as a false start and an object lesson."[95] Hadler does see a crucial role for statins in some rare genetic disorders of cholesterol metabolism and a defensible role in preventing a second heart attack.[96] Were Hadler's "guidelines" followed the very costly resort to statins would be markedly reduced.

Other hazards of taking statins have been identified since Hadler published his concerns in 2004. In 2012, the FDA added some warnings to the safety information on the labels of statins: an increased risk of diabetes by elevating blood sugar; and a risk for cognitive effects such as memory loss and confusion.[97] The FDA downplayed the seriousness of these cognitive side effects, maintaining that the symptoms have generally been reversed by taking people off the use of statins. Furthermore, the FDA issued a ringing endorsement of prescribing statins for preventing heart disease.

However, there is still another side effect, one that is extremely difficult to deal with. The neurologist David Perlmutter refers to a study that found patients receiving statins increased the development of peripheral neuropathy by an amazing 1600%. This disease is often quite debilitating accompanied as it is by burning pain, tingling, numbness, and a loss of sensation in the extremities. As a physician, Perlmutter has found this condition very difficult to treat. It prompts him to advise the medical profession to "reconsider the widespread use of statin drugs."[98] After all, the FDA, in the Physicians Desk Reference, recommends that statins be prescribed only "when dietary means have failed."[99] Patients should be started on diet and exercise programs that have proven to be effecting in reducing heart disease. Perlmutter notes that statins deplete our cells of Co-Q10, a substance that is vital for both brain and heart health, something patients should be told about.

---

[94] Hadler, The Last Well Person, 40–41.
[95] Ibid, 42–43.
[96] Ibid, 43.
[97] Berkrot and Pierson, "FDA Adds Diabetes, Memory Loss to Statins," Union Leader, February 29, 2012, B7.
[98] David Perlmutter and Carol Colman, The Better Brain Book: The Best Tools for Improving Memory and Sharpness and for Preventing Aging of the Brain. (New York: Riverhead Books, 2004), 44. See also, Abramson, JD, Rosenberg, HG, Jewel N and Wright JM, "Should People at Low Risk of Cardiovascular Diesease Take a Statin?" British Medical Journal. 2013: 347:[c6123].
[99] Ibid.

When one reflects upon the very significant, deleterious side effects caused by ingesting statins, and the existence of alternatives, even more effective methods of avoiding heart attacks and strokes, the moral dictum, "first of all do no harm" comes to mind.

4.4.2.6 More Effective Alternatives to Statin Therapy

We have seen that the benefit/harm ratio for statin therapy, certainly for those who do not have heart disease, is not at all favorable. Given the risks of taking statins, it would certainly be worthwhile to know whether there are more effective ways to prevent heart disease that are also free of harmful or undesirable side effects. The good news is that there are and they are much less expensive. The bad news is that they not receive the intense kind of attention and commendations from medical experts, public education campaigns, drug advertisements, and news of breakthroughs in the prevention of heart disease that statins have received since the FDA approved them in 1987. Unfortunately, as Abramson indicates, "the end result is that doctors and patients are being distracted from what the research really shows: physical fitness, smoking cessation, and a healthy diet trump nearly every medical intervention as the best way to keep coronary heart disease at bay."[100]

We begin with studies of physical fitness. A study of 25,000 executives and professional men, published in 1999, collected data from their so-called "executive physical exams" and 10 years later the findings gleaned from these exams were correlated with the deaths that occurred from cardiovascular disease (heart attacks, stroke and blood clots) and all causes to ascertain what factors contributed most to these deaths.[101]

Being among the 20% least physically fit resulted in a far greater health risk than elevated total cholesterol (above 240 mg/dL). Low fitness led to three times as many deaths from cardiovascular disease than did elevated cholesterol for normal weight men. Low fitness among overweight and obese men led to one and a half times as many cardiovascular deaths as elevated cholesterol did. The overall risk of death was even more significant: whereas normal weight men with elevated cholesterol levels had no additional risk, those who were unfit had a 60% higher risk of deaths. Overweight men with elevated cholesterol levels increased their rate of death from all causes by 30% but low fitness increased the death rate by 70%, more than twice as much. Stated in absolute terms, poor physical fitness was associated with seven extra deaths per thousand each year among normal and overweigh men.

Physical fitness has also been shown to be a major factor in protecting women from heart disease and death from all causes. A study begun in the 1970s followed women tested for their fitness for 20 years and found that the least fit had a far

---

[100] Abramson, Overdosed America, 222.
[101] Ibid, 222–223.

greater risk of dying from coronary heart disease and more than twice the overall risk of death.[102]

What does exercise do for those who already have heart disease? In a randomized study of post-heart attack patients, those participating in an exercise program had a statistically significant 27% lower death rate than those in the control group.[103] This benefit is greater than that shown for post-heart attack patients in most of the randomized studies of individuals receiving statins.[104]

Diet, as well as exercise, can more effectively prevent subsequent heart attacks in post-heart attack patients. Those consuming what is called the "Mediterranean Diet" in the Lyon Diet Heart Study, derived a benefit more than two and a half times more than similar patients who are prescribed cholesterol-lowering drugs.[105] This was achieved without lowering total or LDL cholesterol. The overall health benefits of the Mediterranean Diet are also documented in a more recently published study.[106]

However, the expert panel of the National Cholesterol Education Project (NCEP) that issued the 2001 guidelines for lowering cholesterol did not mention the considerably greater benefits of the Mediterranean Diet as compared with similar patients prescribed cholesterol-lowering statins. Furthermore, they omitted any reference to the American heart Association's reaction to the Lyon Diet Study's results. The Heart Association lauded these results as unprecedented reductions in the rates of coronary recurrences, and regards it as short-sighted to ignore the enormous health benefits of the Mediterranean Diet.[107]

The reaction of the NECP panel is part of a more extensive problem: conflicts of interest, five of the 14 experts who wrote the 2001 guidelines including the chair of the panel, had financial ties to manufacturers of statin drugs; four of these five, including the chair of the panel, disclosed financial relationships with the three manufacturers of the best-selling statins.[108] The influences of the pharmaceutical industry range from what studies are supported and published and how they are presented and interpreted: Hadler cites a number of articles that document such influences.[109] Dr. Scott Grundy, who chaired the panel of NECP that created the 2001 guidelines, is quoted by Abramson as telling the Wall Street Journal: "you can have the experts involved, or you could have people who are purists and impartial judges, but then you don't have the expertise."[110] Grundy is reflecting what is true in American medicine presently: many experts are very much financially tied to the medical industry. When Dr. Willett, professor of epidemiology and nutrition at the

---

[102] Ibid, 223.
[103] Ibid.
[104] Ibid.
[105] Ibid, 224.
[106] Estrach R., Ros E., Salas Savado J., et al., "Primary Prevention of Cardiovascular Disease with a Mediterranean Diet," New England Journal of Medicine, 368. April 4, 2013, 1279–1290.
[107] Abramson, Overdosed America, 224.
[108] Ibid, 135.
[109] Hadler, The Last Well Person, 215–216.
[110] Abramson, Overdosed America, 147.

Harvard School of Public Health was asked why a more balanced approach to preventing heart disease is pushed aside by the NECP 2001 guidelines he had this to say: "drug companies are extremely powerful. They put huge efforts into promoting the benefits of their drugs. It's easier for everyone to go in this direction. There's no huge industry promoting smoking cessation or healthy food."[111] In Abramson's own words, "drug companies have much more money to spend promoting the 'scientific evidence' that supports lowering LDL cholesterol with statins than do the flaxseed, canola, olive, soybean, walnut, and vegetable farmers who would benefit from the promotion of the Mediterranean Diet."[112]

What the scientific evidence clearly shows is that interventions, such as smoking cessation, physical fitness, and the Mediterranean Diet, are considerably more efficacious in preventing heart disease and recurring heart attacks than statins. Abramson's version of the Mediterranian Diet is combined with advice against smoking and for exercising, and advice on intake of fats and alcohol.[113] Note that these approaches to preventing and treating heart disease apply just as well to the prevention of type 2 diabetes, osteoporosis, and stroke, and is also helpful in preventing cancer and depression. For these further claims, Abramson has also presented convincing scientific documentation.[114]

We return now to the policy recommended by primary care physicians committed to "Choosing Wisely" to switch to generic versions of statins in order to save money. What is not stated is whether they are following the FDA's recommendation to prescribe statins only after dietary interventions have failed. If so, they would also be greatly curtailing the present extensive resort to statins. Nor does their cost-saving recommendation mention anything about whether they are abiding by the 2001 guidelines of the NCEP that include so many individuals (92,127,000) as candidates for taking statins, and many of them, for being needlessly exposed to their side effects. In any event, the scientific research should incline primary care physicians to reject these guidelines, especially since they call for levels of cholesterol so close to what is average for healthy individuals. They are to be commended for advocating that dosages be titrated in administering statins.

4.4.3 Diabetes Type 2: A Case of Flawed Guidelines and Intervention

Before 1997, a fasting blood sugar over 140 meant you had diabetes. However, in 1997 the Expert Committee on the Diagnosis and Classification of Diabetes Mellitus redefined when a person is said to have diabetes, namely at a fasting blood sugar

---

[111] Ibid.
[112] Ibid, 224–225.
[113] Ibid, 238.
[114] Ibid, 225–238.

over 126.[115] Hence, individuals with a blood sugar reading between 126 and 140 used to be normal but now are diagnosed as diabetics. That resulted in adding one million, six hundred eighty-one people to those regarded as diabetics requiring medical care.[116]

What is the effect of achieving a blood sugar below 140? The National Institute of Health did a study designed to find out whether intensively lowering blood sugar would decrease the risk of having, or dying, from a heart attack or stroke.[117] The trial was conducted with ten thousand patients with diabetes and hence at high risk for the events in question. Five thousand were randomly assigned to receive standard diabetes treatment that would lower their average blood sugar to a more acceptable level but not to a level within the normal range. The other five thousand subjects were randomly assigned to intensive drug therapy that would make their blood sugar normal. Half of these patients did attain the goal with an average blood sugar level below 140. Since the average included measurements obtained immediately after eating when the level tends to be high, it is a safe assumption that their fasting blood sugar levels were quite a bit lower.

The study started in 2003, to last until 2009. However, the study was essentially stopped, early in 2008.[118] The intensive therapy regiment was "changed" because of safety concerns: 5% of patients who received the intensive therapy had died after 3 years as compared with 4% of those who received standard therapy. In examining this study, Welch indicates that is a 25% increase in the risk of death and the researchers did not consider this result to be a statistical fluke: intensive therapy was undoubtedly worse than standard treatment.[119]

Welch explains how making blood sugar reach a normal level could end up killing someone. Blood sugar bounces around and bouncing around normal the level can bounce too low, increasing one's risk of death. The investigators professed not to know what caused the increased mortality. Welch doesn't buy that! His view is that, had the trial yielded a mortality benefit it would have been ascribed to what was being tried, namely intensive control of blood sugar. Therefore, since the trial yielded a mortality harm that should be ascribed to the intensive control of blood sugar—the very point of engaging in a randomized trial.[120]

As far as Welch is concerned, the lesson to be drawn from this research is that, "if it's not good to make diabetics have near normal blood sugars, then it's not good to label those with near normal blood sugars as diabetics."[121] It's not good, Welch

---

[115] "Report of the Expert Committee on the Diagnosis and Classification of Diabetes Melitis," Diabetes Care 20, 1997, 1183.

[116] I.M. Schwartz and S. Woloshin, "Changing Disease Definitions: Implications on Disease Prevalence," Effective Clinical Pracice 2, 1999, 76–85.

[117] Action to Control Cardiovascular Risk in Diabetes Study Group, "Effects of Intensive Glucose Lowering in Type 2 Diabetes," New England Journal of Medicine 358, 2008, 2545–2549.

[118] Welch, Overdiagnosed, 19.

[119] Ibid.

[120] Ibid, 19–20.

[121] Ibid 20.

argues, because those so labeled will be treated "and people with mild blood sugar elevations are least likely to gain from treatment—and arguably most likely to be harmed...."[122]

The present guidelines that lead to overdiagnosing and overtreating diabetes 2 should be corrected so as to avoid these practices. Why? First of all, the principle of first of all do no harm should be followed. After all, the harms to be avoided, include increasing an individual's risk of death. Secondly, avoiding these overdiagnoses and unnecessary treatments will greatly reduce the high costs now spent on medical care for diabetics. Thirdly, it will contribute to well being. Receiving the diagnosis that one has diabetes can be rather frightening, certainly troubling. In addition to correcting the faulty guidelines, the scientifically sound recommendations for preventing diabetes, as well as the risk of heart disease associated with diabetes, are a good diet, exercise, and avoiding smoking and not being overweight.

As we have noted previously, the independence of experts who set the parameters of conditions, in this instance of when someone is to be diagnoses as diabetic, is rendered questionable by their ties to pharmaceutical companies. The head of the diabetes panel was a paid consultant to the following manufacturers of diabetes drugs: Aventis Pharmaceuticals, Bristol-Myers Squibb, Eli Lilly, GlaxoSmithKline, Novartis, Merck, and Pfizer.[123] Welch allows for the possibility that many of these experts may be conscientiously seeking to benefit people they believe should receive therapeutic interventions. However, with so much money at stake, he alleges that experts can be led to overestimate the benefits and overlook the harms of overdiagnosis. Hence, since such decisions affect so many individuals, they should not be tainted by the businesses that will gain from them.[124] One should also consider the cost of such decisions to taxpayers and the very solvency of government-funded health care, particularly since curbing overdiagnoses also increases the quality of medical care. In the arena of overdiagnosis, less is better than more.

4.4.4 Additional Overdiagnosed and Overtreated Physical Conditions

So far in this chapter we have provided some detailed examples of overdiagnosing and overtreating taking place in medicine as currently practiced. When we speak of overtreating we are referring to treatments that cannot be adequately defended on scientific and moral grounds. The examples analyzed were covered so extensively in order to present the scientific research that documents more effective, less costly interventions that, at the same time, are free of harmful side effects of the presently dominant treatments being offered. After all, no physicians would or should change

---

[122] Ibid.
[123] Ibid, 24.
[124] Ibid.

4.4 Overdiagnosing and Overmedicating: Practices that Need to Be Curbed

what they conscientiously do to help their patients unless they are made aware of soundly scientific and moral reasons for changing: hence the highly detailed accounts heretofore.

However, it would require a very thick volume indeed to provide the reader with additional, similarly detailed analyses of other instances of overdiagnosing and overtreating. Instead what now will follow are very brief descriptions of a few examples of such instances appropriately footnoted so the interested reader can pursue the basis for reducing and/or changing the practices in question. These additional examples are intended to give some, though still incomplete, further indication of the enormous amount of money that could be saved while greatly improving the quality of medical care. Consider also that by the time this books is published, the Choosing Wisely campaign may have recommended some of these very same changes suggested by the scientific research we are documenting, and may have also altered practices we have not covered. For example, there is an urgent need to deal with the preventable and predictable deaths that result from allegedly unnecessary surgery, interventions we will have only partially singled out for attention. We turn now to our brief references to further examples of overdiagnosing and overtreating.

4.4.4.1 Hypertension

Changing what was considered to be a normal level of blood pressure into an abnormal level, added 13, 490,000 people regarded as in need of treatment with drugs. This turns a great number of individuals into candidates for being overdiagnosed and overtreated, with the possible exception of the elderly who can benefit somewhat from mild drug treatments. However, they can recieve the same benefit from weight reduction and/or reduced salt intake as applicable in any given case.[125]

4.4.4.2 Prostate Cancer

Screening as now practiced results in considerable overdiagnosis and overtreatment: it is not decreasing mortality and the harms of treatment are very serious and extensive.[126] To cite only one example of the ineffectiveness of screening and treatments, much higher rates of screening, doing biopsies, and treating with radiation and prostectomies, men 65–79 in Seattle as compared with those in Connecticut, yielded no difference in the likelihood of dying from prostate cancer.[127]

---

[125] Ibid, 20, 23–24 and 27–28. See also Hadler, The Last Well Person, 54–58.
[126] Welch, Overdiagnosed, 45–60. See also Hadler, The Last Well Person, 92–100.
[127] Hadler, The Last Well Person, 98–99.

4.4.4.3 Colorectal Screening

What is clearly being overdiagnosed are polyps considered to be precancerous, and detected and removed by the use of colonoscopies. Some colonoscopies lead to perforations that require surgery; others result in heavy bleeding that require surgery to quell in some cases. These harms are generally not being conveyed in the widespread promotion of colonoscopies. Nor is there sufficient awareness that death from all causes is not reduced by colorectal screening.[128]

4.4.4.4 Thyroid Cancer

From 1975 to 2005 new diagnoses of thyroid cancer more than doubled. However, the death rate from thyroid cancer was unchanged. This inevitably increased treatments that are unnecessary, subjecting some to harmful side effects, and, in all cases, to being medicated for the rest of their lives. Certainly individuals with symptoms should be screened, diagnosed, and treated if cancer is discovered. However, people who are well may indeed have asymptomatic tiny cancers that are revealed by CT scans or felt in the throat as lumps. Based on their research, some have suggested that it is so common to have tiny cancers in one's thyroid that this should be considered normal.[129]

4.4.4.5 Cardiac Interventions

The United States does three and a half times as many coronary angioplasties and coronary artery bypass surgeries on a per-person basis for patients of all ages than other industrialized nations. Yet the United States has the third highest death rates from coronary artery disease among ten wealthy industrialized nations. Between 1998 and 2000 the in-hospital death rates for heart attack patients remained virtually unchanged despite increased interventions. As compared to Canada, patients 65 and older are five times more likely to receive balloon angioplasty or coronary artery bypass surgery. Nevertheless, as many patients treated in the United States die 1 year after their heart attacks as those treated in Canada.[130]

Coronary artery bypass surgery is not only ineffective but it is also quite harmful. There are 2–8% who die on the table or post-operatively; 50% who suffer emotional distress in the first 6 months; 40% who have memory loss; and many who can no longer work or describe themselves as well and enjoying their lives. All this occurs while the chance of survival is not improved.[131] There is evidence that catheterization after a heart attack is also harmful and devoid of benefits. It results in increasing

---

[128] Welch, Overdiagnosed, 69,71 and 101. See also Hadler, The Last Well Person, 64–76.
[129] Welch, Overdiagnosed, 62–66.
[130] Abramson, Overdosed America, 171–173.
[131] Hadler, The Last Well Person, 24–25.

mortality, heart pain (angina), and decreased ability to perform tasks that require moderate exertion.[132]

4.4.4.6 Infant Intensive Care

In the United States, much of the country has approximately twice as many intensive care services as are needed to obtain optimal survival rates. Neonatal mortality is not further reduced by intensive care once a basic level of such care is available for newborns. As compared with Australia, Canada, and the United Kingdom, the United States has twice as many neonatologists and neonatal intensive care beds for each baby born. Still, survival rates for babies of equivalent birth-weight are no better in the United States.[133]

4.4.4.7 Monitoring Pregnant Women

Monitoring the electric impulses of a baby's heart in utero during labor leads to a significantly greater number of emergency c-sections for those so monitored than for those who have not been monitored. Emergency c-sections are more risky than planned c-sections. The only benefit is the prevention of one seizure in infants per 130 c-sections. It is difficult to compare the risk of seizures to the risk of c-sections. In 1996, the U.S. Preventative Services Task Force that reviews screening, recommended against such monitoring on a routine basis. However, as of 2013, they did not renew this recommendation because it was extensively ignored. Women should at least be made aware of the high risks of emergency c-sections and the very rare possibility of any seizures in their infants.[134]

Women are quite commonly screened with the use of ultrasound (sonograms) that can detect abnormalities associated with trisomy syndrome. Down syndrome, involving mental retardation, is the most familiar trysome syndrome. Trisomy syndromes are rare: they occur in three out of every thousand live births. However, the abnormalities tied to these syndromes are found in 10% of normal fetuses. That means that at most only three of every hundred fetuses said to have abnormalities will actually have trisomy. Those discovered to have abnormalities will be further tested with tests that are not definitive. A review of fifty-six studies of these abnormalities led the authors to conclude that they are not reliable enough indicators of trisomy for purposes of clinical practice: ultrasound leads to more miscarriages than diagnoses of trisomy syndrome. As in the case of fetal monitoring, the U.S. Preventive Services Task Force recommended against routine use of obstetrical ultrasound in 1996 and subsequently did not renew their advisory because they did not expect it

---

[132] Abramson, Overdosed America, 171–172.
[133] Ibid, 174.
[134] Welch, Overdiagnosed, 105–107.

to curb its use.[135] This practice bears further scrutiny and, at the very least, providing thoroughly informed consent for pregnant women.

4.5 Briefly Assessing Where WE'VE Been and Where We're Going

The aim of this book is to make a case for achieving justice in the provision of medical care. To that end, we began by conceptualizing justice. Given what justice demands, everyone's medical needs should be met regardless of how much, if anything, someone can pay for such services. Because there are those who regard justice as satisfied if everyone has access to a basic minimum of medical care, we devoted Chap. 3 to revealing the flows in that kind of thinking about justice. Those flawed conceptions of justice implicitly sanction the rationing of medical care, explicitly so in the one case examined. That requires us to answer the question: Is rationing avoidable or is it not?

In this fourth chapter, we began by acknowledging that continuous rising costs of medical care in the United States are unsustainable. To avoid rationing, something has to change. As it turns out, the medical community is beginning to recognize, that on scientific and moral grounds, a great many very expensive interventions ought to be eliminated or curbed, as the case may be. The rapidly expanding Choosing Wisely movement attests to that. However, this movement, though expanding, has a long way to go. Scientifically and morally, a great deal more very expensive and harmful medical practices need to be curtailed, and in some instances ended totally, for the sake of making way for the proven and available more effective, less costly, and mostly harm-free alternative interventions such as diet and exercise.

The reader needs to understand that the examples discussed in this chapter and the next are but a limited set of very expensive practices that were identified and argued to be morally unjustifiable as they now stand. The emphasis for selecting them was to highlight the importance of preventing harm in instances where this is possible. We want it known that the quality of care can be greatly increased and costs greatly decreased when the demands of justice are more nearly satisfied.

As we shall see in the next chapter, the quality of care for the mentally ill is being severely undermined by certain harmful and unjustifiably unnecessary interventions. We have noted how relations between medical experts and the pharmaceutical industry work to set standards that favor ever more treatment. Those conflicts of interest also strongly determine ever expanding drug usage for treated those diagnosed as mentally ill. What is more, pharmaceutical companies more directly create and market their products for behavior and feelings they have identified as abnormal. And they have done so very profitably.

---

[135] Ibid, 107–113.

Chapter 5
Overdiagnosing, Overtreating, and Overmedicalizing Behavior and Feelings

Abstract This chapter continues to pursue and document unjustifiable medical practices that need to be curbed as much as possible so as to eliminate this source of unsustainable costs. In this case, the source is the enormous diagnostic inflation of mental disorders. Diagnostic inflation treats normal behavior and feelings as mental disorders and thus medicalizes them. This comes about in two ways: failing to abide by the criteria for mental illness, often fueled by deceptive drug ads, and failing in the process of diagnosing by psychiatrists and other medical practitioners: for example, diagnosing on the basis of one visit and not taking the time to learn if a troubled individual is having normal feelings or behaving normally.

In the United States, mental illness is increasingly being overdiagnosed. Currently about half of the American population are deemed by one source to be qualified for a lifetime mental illness.[1] There are those who consider this to be an underestimation of the prevalence of mental illness in the U.S. One prospective study projects that "by the age of thirty-two 50 percent of the general population had qualified for anxiety disorder; more than 40 percent for mood disorder; and more than 30 percent for substance dependence."[2] Not to be outdone, another study had 80% of adults at age twenty-one meeting the criteria for a mental disorder.[3]

There is ample evidence of diagnostic inflation of mental disorders. The noted psychiatrist, Allen Frances, has us consider the enormous increases in such diagnoses in the past 15 years:

> Child bipolar disorder increased by a miraculous fortyfold; autism by a whopping twentyfold; attention deficit/hyperactivity has tripled; and adult bipolar disorder doubled. Whenever rates skyrocket some portion of the rise represents previously missed true cases—people who really need the diagnosis and the treatment that follows from it. But

---

[1] Allen Frances, *Saving Normal: an Insider's Revolt Against Out of Control Psychiatric Diagnosis, DSM-5, Big Pharma and the Medicalization of Ordinary Life*, (New York: HarperCollins Publishers, 2013), 103.
[2] Ibid, 104.
[3] Ibid.

more accurate diagnosis cannot explain why so many people, especially kids, suddenly seem to be getting sick.[4]

With so much overdiagnosing, the use of psychotropic drugs (those acting on the mind) has greatly accelerated. Antidepressants, stimulants, anxiety agents, sleeping pills and pain medications are presently among the best selling drugs for the drug companies. Annually, 300 million prescriptions for psychiatric drugs are issued in the U.S.: antipsychotics reap 18 billion dollars a year; antidepressants bring in 12 bullion dollars a year; stimulants yield 8 billion dollars a year, up from 50 million dollars 15 years ago.[5]

Frances regards it as truly puzzling that now more than three million Americans are being issued antipsychotics, a number that grows at a rate of 20% per year, and that in 10 years the number of prescriptions has doubled to 54 million.[6] This puzzles Frances because antipsychotics have dangerous side effects and the indications for them are narrowly confined to the disabling symptoms of schizophrenia and bipolar disorder. Yet drug companies promote the use of antipsychotics for those who have "trouble sleeping, or run-of-the-mill anxiety or depression, or irritability, or eccentricity, or the temper tantrums of youth, or the crankiness of old age."[7]

Along with the prescriptions for antipsychotics, off-label uses of drugs have doubled in 10 years. This happens despite the fines sustained by drug companies: the profits attained dwarf them, rendering the fines utterly ineffective in halting such practices. Frances notes as examples 2.4 billion a year spent on advertising Abilify and Seroquel, drugs he calls "very so-so and not-so-safe." As a result, these drugs placed fifth and sixth among all medicines being sold in the U.S. Furthermore, the drug manufacturers work diligently to influence primary care physicians and have them dispensing antipsychotics inappropriately for 20% of all their anxiety disorder patients. Frances considers this "massive misuse of marketing might over common sense and good medical practice.[8]

Frances is truly alarmed by these excessive and dangerous practices. They prompted him to publish his book <u>Saving Normal</u> that bears the subtitle "an insider's revolt against out-of-control psychiatric diagnosis, DSM-5, big pharma, and the medicalization of ordinary life."[9] The DSM is the <u>Diagnostic and Statistical Manual of Mental Disorders</u>, the leading guide for the practice of psychiatry. Frances headed the authors who issued DSM-IV so he is well qualified to critique the rather loose criteria and implausible new disease categories found in DSM-5.

Frances is not alone in observing and documenting what he refers to as the "medicalization of ordinary life." Peter Conrad is among a number of sociologists who have studied the phenomenon of medicalization. Having begun this scholarly pur-

---

[4] Ibid.
[5] Ibid, 104–105.
[6] Ibid, 105.
[7] Ibid.
[8] Ibid.
[9] Ibid, title page.

suit in the 1970s, he has had the opportunity to document the phenominal growth of medicalization.
Conrad alerts us to the fact that:

> Numerous studies have emphasized how medicalization has transformed the normal into the pathological and how medical ideologies, interventions, and therapies have reset and controlled the borders of acceptable behavior, bodies and states of being.[10]

Frances dramatically describes his book as an effort to save us all from these very "powerful forces trying to convince us that we are all sick."[11]

The term medicalization refers to a process by means of which a problem regarded as non-medical comes to be defined and treated as a medical problem, usually characterized as an illness or disorder. To call a problem medical or medicalized is to assert that it is described by using medical language, understood by adopting a medical framework or treated by means of medical interventions.[12] Medicalization as a process can and does lead to identifying illnesses or disorders that are genuinely medical and benefit from being treated medically. However, there are increasing instances of medicalizing that transform the normal into the pathological and that cannot be scientifically and ethically justified. This is what we mean when we speak of overmedicalization. We turn now to detailed documentation and analyses of instances of overdiagnosing and overmedicalizing, and the unjustifiable treatments that result from these activities.

5.1 Diagnostic Inflation and the Faulty Treatment of Mental Illnesses

5.1.1 Depression

To begin with, the psychiatric diagnostic manual (DSM) too loosely defines a major depressive disorder. The definition is also arbitrary: having five symptoms for only 2 weeks duration.[13] Even if one were to disagree with this assessment with the psychiatrist Frances, anti-depressants are used in circumstances questionably true to the manual's diagnostic criteria and, in many instances, clearly violate them.

Primary care physicians dispense 80% of the prescriptions for antidepressants and almost three quarters of the 11% of the population taking them have no symptoms of depression.[14] Although Frances acknowledges that some primary care physicians are well informed, he regards it as inappropriate to have depression diagnosed

---

[10] Peter Conrad, The Medicalization of Society: On the Transformation of Human Conditions Into Treatable Disorders, (Baltimore: The Johns Hopkins University Press, 2007), 13.
[11] Frances, Saving Normal, xx.
[12] Conrad, The Medicalization of Society, 4–5.
[13] Frances, Saving Normal, 23–24.
[14] Ibid, 100–101.

and treated by primary care physicians: that should be the exclusive purview of those with adequate psychiatric training.[15]

To explain why primary care physicians are misusing antidepressants, Frances first documents the healing power of the placebo effect.[16] As he notes, "being able to enlist the confidence and hope of the sick patient has always been and still is the most essential skill in a great shaman or a great modern doctor.[17]

Drug companies are making a great deal of money (12 billion dollars annually) by exploiting the power and ubiquity of the placebo effect. To get results for antidepressants, and other psychotropic drugs for that matter, is to treat people who do not need them and would naturally get better in time on their own. The clever marketing trick was to persuade doctors to treat patients who were not actually sick, and at the same time, to sway normal people to regard themselves as sick. So drug manufacturers advertised extensively throughout the world on behalf of what are essentially expensive placebos.[18]

Many of the patients, somewhere between 25% and 50%, came to the primary care physician, at least in part, because they are somewhat emotionally distressed.[19] Most of the patients these physicians treat have only a mild disorder, the kind most likely to benefit from the placebo effect. Once they recover, they will usually wrongly think the medicine is what helped them. That may lead them to feel that they ought to stay on the medication for prolonged periods. Patients and physicians have been subjected to misleading marketing by drug companies. Indeed, a great deal of the primary care physicians "education" in psychiatry is offered by drug company sales persons who also come bearing free samples. The typically underpaid and overworked primary care physician, confronted by patients who likewise have been "taught" in drug advertisements to seek particular drugs to cope with their particular emotional state, will find it convenient and appropriate to prescribe, or give out a sample of, the drug being sought by the patient.

Abramson, like Frances, is quite disturbed that drug ads bring him patients requesting pills to keep them well, and not inquiring into the changes in their lives that would benefit them far more. Too often patients turn their visits into a contest of wills, an unproductive alternative to engaging in helpful dialogue focused on their health risks and habits.[20]

---

[15] Ibid, 102–103.
[16] Ibid, 97–99.
[17] Ibid, 99.
[18] Ibid, 100.
[19] Ibid, 102.
[20] John Abramson, <u>Overdosed America: The Broken Promise of American Medicine.</u>, (New York: HarperCollins Publishers, 2004), 10.

5.1.1.1 Evidence for the Relative Ineffectiveness and Harmfulness of Antidepressant Drugs

The new antidepressant drugs received very high marks from respected primary care physicians in 2011. They ranked them as the eight most important medical innovations of the last 25 years among thirty medical innovations they were asked to rank.[21] They did this on the basis of the scientific evidence available to them in reputable journals that portrayed these drugs as beneficial. On what grounds, then, can the effectiveness of these antidepressants be questioned?

First of all, not all the studies done on these antidepressants were published. The Swedish Drug Authority is one of the sources we have for exposing what they call the "biased evidence" one receives from published recommendations of a specific drug.[22] With respect to the applications this Drug Authority received for approval of five new antidepressants, half of the studies, 21 of 42, showed that antidepressants were no more effective than placebos. Yet only six of these 21 negative or inconclusive studies were published, while 19 of the positive studies were published in 22 articles, three being published twice. As Abramson notes, doctors conscientiously weighing evidence that they could find in the medical journals, would be led to believe that antidepressants were effective.[23]

Researchers in the U.S., well aware of this publication bias, obtained through the Freedom of Information Act the results from all of the studies, both published and unpublished, that the FDA had reviewed for the purposes of approving seven new antidepressants (Prozac, Zoloft, Paxil, Effexor, Serzone, Remeron, and Wellbutrin SR).[24] When all of the evidence was considered, the new antidepressants were not more effect than older antidepressants and not much more effective than placebos: symptoms of depression improved by 30.9% for individuals who took placebos; by 40.2% for individuals who took the new antidepressants, and by 41.7% for individuals who took the older antidepressants. The data also revealed that in nine out of ten studies, the new drugs are no more effective than placebos for patients with mild depression, the majority of depressed patients primary care patients see.

The new antidepressants are more dangerous than the older tricylic antidepressants in so far as they generate a higher rate of death from suicide.[25] What is more, individuals receiving the new antidepressants successfully commit suicide at twice the rate as those receiving placebos: for every thousand people with depression treated with a new antidepressant drug, 4.6 more committed suicide than those given a placebo.

A study published in 2001 found that Paxil, a new antidepressant, is effective and safe for depressed adolescents: depressed adolescents improved significantly more

---

[21] Ibid, 114–115.
[22] Ibid, 115.
[23] Ibid.
[24] Ibid, 115–116.
[25] Ibid, 116.

when treated with Paxil than when treated with a placebo.²⁶ This was all the "evidence" doctors had available to them: it turned out to be another example of publication bias. A review by British drug authorities of all nine studies, only one of which had been published revealed quite different overall results: those taking Paxil were no less depressed than those on placebos.²⁷ Furthermore, the instances of instability including suicidal thoughts, was twice as high for the adolescents given Paxil as compared with the adolescents given placebos (3.2% versus 1.5%). That prompted the United Kingdom Medicine and Healthcare Products Regulatory Agency to advise that Paxil and other SSRI (selective serotonin uptake inhibitors) drugs, the new antidepressants, not be prescribed for patients under the age of 18.²⁸

As compared with older antidepressants, the new antidepressants have about 12% fewer side effects. However, Abramson warns against the routine use of the new antidepressants for all depressed patients considering the increased risk of suicide and relative ineffectiveness: the selective publication of commercially sponsored research leads doctors to believe that these newer drugs are more effective and less dangerous when the opposite is true. And they are costly. These new antidepressants were the best selling class of drugs in the U.S. between 1999 and 2001 and ranked third in 2002 and 2003. Americans spent 12.5 billion dollars on them in 2001.²⁹ They have been averaging 12 billion dollars annually in sales, as reported earlier above.

There is another study testing Zoloft, one of the new antidepressants, that cast doubt on these drugs as treatments of choice for depression.³⁰ Patients diagnosed as having major depression were randomly assigned to three groups: one to receive Zoloft; one to receive three exercise sessions a week; and one to receive both Zoloft and exercise; all for four months. After four months, depression was significantly reduced in all three groups. However, six months after the treatment ended, only 8% of the individuals in the exercise group became depressed again, while 38% of the individuals treated with Zoloft only, and 31% of those treated with both Zoloft and exercise suffered relapses.

Abramson understands these differing outcomes as showing that: "short-term treatment with an antidepressant medication relieves symptoms but appears to decrease the likelihood of patients making the positive life changes necessary to prevent symptoms from recurring.³¹ Indeed, Abramson adds that the study suggests that some depression might well be called "exercise-deficiency disease." Certainly this study adds more substance to Abramson's contention, and that of Frances that antidepressants should not be routinely prescribed for all depressed patients.

Having noted that for most people with mild or transient symptoms, the SSRIs (new antidepressants) amount to "nothing more than very expensive and potentially

---

²⁶ Ibid, 116–117.
²⁷ Ibid, 117.
²⁸ Ibid.
²⁹ Ibid.
³⁰ Ibid, 233.
³¹ Ibid.

harmful placebos,"[32] Frances has the following suggestions: "people should have more faith in the remarkable healing powers of time, natural resilience, exercise, family and social support, and psychotherapy—and much less faith in chemical imbalances and pills."[33] Sadness should not be considered a disease. It is harmful to label people. Being labeled diminishes one's sense of worth. It can act as a self-fulfilling prophecy, making one labeled as mentally ill feel and act sick, and prompt others to treat you as one who is sick. The sick role helps those in genuine need of care, but as Frances observes, "the sick role can be extremely destructive when it reduces expectations, truncates ambitions, and results in a loss of personal responsibility."[34]

The diminished resort to psychotherapy is proving costly now and will even more so in the future. The percentage of visits to psychiatrists who engage in psychotherapy has dropped from 44% in 1996 to 1997, to 29% in 2004–2005. Yet it is as effective as drugs in treating those who have mildly to moderately severe kinds of problems. Its effects are more enduring, reducing costs in the long run, making it cheaper than long-term medication. Unfortunately, psychotherapy is not as well compensated as the shorter sessions that simply manage medications.

Japan has caught on to the advantages of psychotherapy both in reducing costs and in putting patients in charge by bringing about new coping skills and attitudes toward life. In this way Japan breaks up the drug monopoly with the more effective and less costly resort to cognitive therapy.[35] Unless alternatives to antidepressants and psychotropic medications more generally become a standard practice in the U.S., the Affordable Care Act will substantially increase the costs of medical care because of its extension of health care coverage and its requirement that insurance cover comprehensive care for mental illness.[36]

5.1.2 Attention Deficit Disorder (ADHD)

In the span of 15 years, the number of individuals diagnosed as having ADHD has tripled.[37] For Frances, this is a "false epidemic."[38] Such a precipitous rise in such a short time cannot be explained by regarding this as a matter of finding previously missed cases. Some of that is surely involved but that cannot begin to account for so many, so suddenly becoming ill.[39]

---

[32] Frances, Saving Normal, 157.
[33] Ibid.
[34] Ibid, 109.
[35] Ibid, 108.
[36] Ibid, 112.
[37] Ibid, 104.
[38] Ibid, 75.
[39] Ibid, 104.

What then explains this rampant overdiagnosing of ADHD? Frances identifies the following six conditions:

> Wording changes in DSM-IV; heavy drug marketing to doctors and advertising to the general public; extensive media coverage; pressure from harried parents and teachers to control unruly children; extra time on tests and extra school services for those with an ADHD diagnosis; and finally, the widespread misuse of prescription stimulants for general performance enhancement and recreation.[40]

An amazing 10% of children are now diagnosed as having ADHD as all manner of classroom disruption has become medicalized. What is being labeled as a medical disorder are attentional and behavioral problems once viewed as part of normal life and individual differences.

There is evidence supporting one major source for these excessive diagnoses. When a child is born is a very powerful predictor of whether a given child will receive the diagnosis of ADHD. Because January 1 is taken as the cut off point for grade assignment, boys born in January have a 70% higher risk than those born in December to be diagnosed as having ADHD. The birthday effect is about as strong for girls. Thus, Frances concludes: "we have turned being immature because of being young into a disease treated with a pill."[41]

In a thoroughly documented account, the sociologist Peter Conrad corroborates the first five of the six sources Frances identified as driving up the expansion of diagnosing ADHD.[42] However, he does not agree with Frances that the DSM guide for the practice of psychiatry is essentially a scientific endeavor. Rather, that document is a "mix of social values, political compromise, scientific evidence, and material for insurance forms."[43] After all, DSM-IV contains almost four hundred distinct medically diagnosable categories of mental illness.[44] That's a lot of medicalizing!

It is true that Frances purports that he and his colleagues were rigorous in seeking scientific backing for their decisions in assembling the DSM-IV.[45] Yet he views the DSM as having become "the object of undue worship as the 'bible' of psychiatry."[46] He regrets that the criteria for ADHD were loosened, contributing to overdiagnosing it, especially in the light of the flood of drug company direct to consumer ads.

---

[40] Ibid, Ibid, 141.

[41] Ibid, 141–142.

[42] Conrad, The Medicalization of Society, 41–69. This account races the expansion of diagnosing ADHD children into finding it in adults.

[43] Ibid, 48.

[44] Ibid, 49.

[45] Frances, Saving Normal, xiii.

[46] Ibid, xii.

5.1.2.1 The Spread of ADHD Diagnoses to Include Adults

Frances is incredulous and disturbed, that despite the false epidemic of ADHD already very much in evidence, DSM-5 has made it so easy to diagnose ADHD in adults that a new false epidemic may very well come about. After all, "attention problems are nonspecific and extremely common among normal adults and in those suffering from any of the other mental disorders."[47] Relying on DSM-5 will mislabel many normal individuals who are dissatisfied about their ability to concentrate and complete their work, particularly if they experience boredom or dislike the work they do. It will also tend to misdiagnose those who actually suffer from one of a number of mental disorders. Receiving stimulants inappropriately can worsen any existing psychiatrically treatable condition they may have.[48]

ADD is even now too easy to diagnose. There are many among perfectionists and those over 50 years old who perceive themselves as having difficulty with attention and concentration. In any event, their symptoms are largely subjective, based as they are on fallible self perceptions. The lowered requirements in DSM-5 will pick up many adults who wish to be sharper and do not have any kinds of problems that are serious enough to regard them as mentally ill. Misdiagnosis will also be common for college students, those who have demanding jobs, and even for those who have trouble staying awake, for example, long-haul truck drivers.[49] Frances strongly urges that the criteria for diagnosing ADHD in adults be rendered more rigorous. As he notes "thirty percent of college students cannot suddenly have developed ADHD."[50]

Peter Conrad extensively documents the ever increasing promotion of ADHD as a mental disorder of adults in articles by professionals and the media, books for popular consumption, and television news reports and news shows.[51] A particularly influential book, Driven to Distraction by two Harvard Medical School psychiatrists was a best seller. The authors, Edward Hallowell and John Rotey, developed their own criteria for recognizing attention deficit disorder without hyperactivity. Using their set of criteria, and a hundred question test that Conrad regards as elusive, they urge readers to evaluate themselves for attention deficit disorder (ADD), and on that basis seek professional help for a possible diagnosis if they deem it is prudent or necessary.[52]

There are also various organizations that serve as support groups for those with ADHD. The largest among them is CHADD (Children and Adults with Attention Deficit/Hyperactivity Disorder).[53] CHADD views ADHD as a lifespan disorder, one effecting individuals at all ages. CHADD portrays it as a neurological disorder, not

---

[47] Ibid, 184.
[48] Ibid.
[49] Ibid, 184–186.
[50] Ibid, 186.
[51] Conrad, The Medicalization of Society, 53–55.
[52] Ibid, 55.
[53] Ibid, 55–56.

as a psychiatric or behavioral disorder, thereby bolstering its stake in disability entitlements. CHADD promotes the existence of adult ADHD, advocates for those who think they have it, and legitimates the disorder for them.[54]

Another substantial push for diagnosing and treating ADHD comes from the pharmaceutical firm Ciba-Geigy that manufactures Ritalin, reported in 1999 to be prescribed the most for treating ADHD.[55] Framing hyperactivity, and later ADHD, has occupied this company for a long time. As early as 1971, Ritalin yielded 15% of Ciba's gross profits. Once ADHD was defined as a lifetime disorder, children and adults can be kept on medication for an indefinite period. The treatment of adults contributed to what was an eightfold increase in the U.S. use of stimulants over the decade preceding a 1999 review of its use. The influence of Ciba has also been exerted by contributing quite considerable sums of money to CHADD, joining other pharmaceutical companies in supplying CHADD with 22% of its nearly 4.5 million dollars revenue in 2004–2005, for example.

We can see why Frances is so concerned that DSM-5 has loosened criteria for ADHD. Drug companies, advisory groups, and some notable psychiatrists were already stretching the bases for diagnosing ADHD given the criteria developed in DSM-IV. Frances wishes he had been able to anticipate all of this and had tightened the criteria that appeared in DSM-IV under his leadership. After all, what Frances refers to as the false epidemic of ADHD is not only financially costly and wasteful, but it harms those who are misdiagnosed and unjustifiably treated with drugs.

5.1.2.2 The Harmful Effects of Overdiagnosing and Overtreating ADHD

Though Frances indicates that stimulants can be generally safe and effective, that is only true when someone is properly diagnosed and supervised. Those who are overdiagnosed and unjustifiably treated with drugs are subjected unnecessarily to the risks of serious side effects.[56] These harmful side effects include insomnia, loss of appetite, irritability heart rhythm problems, and a variety of psychiatric symptoms.[57] ADHD drugs have also been cited for excess deaths.[58]

There are further deleterious effects that result from the ever expanding misdiagnosis of ADHD. In adults, the symptoms of ADHD are manifested in how they perform, whether they are underachievers or underperformers. Since those problems with performance are regarded as brought about by an inborn neurological condition, they have a medical explanation for their underperformance that allows them to reevaluate their past behavior, and by shifting responsibility, reduce self-blame for whatever is problematic about their past behavior.[59]

---

[54] Ibid, 56.
[55] Ibid, 56–57.
[56] Frances, Saving Normal, 185.
[57] Ibid, 143.
[58] Conrad, The Medicalization of Society, 129.
[59] Ibid, 64–65. See Kate Kelly and Peggy Ramundo, You Mean I'm Not Lazy, Stupid or Crazy? A Self-help Book for Adults with Attention Deficit Disorder, (Cincinnati: Tyrell and [Severn], 1993) for an example of reduced self-blame writ large.

5.1 Diagnostic Inflation and the Faulty Treatment of Mental Illnesses

The loss of personal responsibility is no small problem. In a section of his book entitled "the Power to Label is the Power to Destroy," Frances discusses what he refers to as "secondary harms" that result from being labeled as one who is mentally ill.[60] Labels can create self-fulfilling prophecies. Having been told you are sick leads one to feel and act sick, and having others treat you as sick. Adopting the sick role is extremely useful when one is actually sick and in need of rest and care. However, as Frances points out, "the sick role can be extremely destructive when it reduces expectations, truncates ambitions, and results in loss of personal responsibility." Furthermore, "when a society allows the overdiagnosis of a significant proportion of its individuals as 'sick' it becomes an artificially 'sick' society rather than a resilient one."[61]

Conrad, publishing in 2007, discusses how ADHD has gone from a "'medical excuse' for life's problems" to being eligible for the benefits provided by the Americans with Disabilities Act (ADA), and also for special accommodations in the workplace and adult education.[62] Thus one has to add to the harmful side effects of wrongfully prescribed drugs, the cost of lost work productivity and of what Frances call "chronic vocational invalidism" of those overdiagnosed individuals deemed eligible for the financial benefits provided by ADA.[63]

Extending a greatly inflated rate of diagnosing ADHD to adults is very costly indeed. Not only are there the costs of drugs that are so largely unneeded and the costs of visits to the doctors for prescriptions but also the costs of an inflated and growing number of individuals who generate costly debate with psychiatrists on both sides arguing whether someone is to be declared insane or culpable for a given crime. These "endless and futile debates" at only one death penalty hearing can cost 5 million dollars, and a year for someone inappropriately sent to a psychiatric hospital "costs more than a year at Harvard."[64] Frances is also concerned that with the loss of personal responsibility that accompanies a diagnosis of mental illness, diagnostic inflation urgently needs to be reduced in the courts. Why? Because it "turns into mad" and, as he asserts: "Bad should usually trump mad."[65]

These are only some of the enormous costs and harms that could be avoided if ADHD were properly diagnosed. After all, recall that the numer diagnosed as having ADHD has tripled in the past 15 years. Consider also, that many of the costs we have described apply to the inflated diagnoses of mental illnesses more generally and apply also in the cases of purely invented categories of mental disorders, some of which we will be identifying later in this chapter.

---

[60] Frances, Saving Normal, 109.
[61] Ibid, 109 & 110.
[62] Conrad, The Medicalization of Society, 65–66.
[63] Frances, Saving Normal, 111.
[64] Ibid, 112.
[65] Ibid, 110–111.

5.1.3 Bipolar Disorder

This disorder is yet another instance of a false epidemic. As the one who chaired DSM-IV, the guide to psychiatric diagnoses, Frances regrets that the manual did not do more to sound a warning about the risks of overdiagnosis and provide tips as to how to avoid it. He and his colleagues did not anticipate the sheer power of pharmaceutical companies and hence did not do enough to reduce the ability of drug companies to sell sickness.[66] As Frances observes, the last 15 years teaches a definite lesson: the DSM alone does not set standards. In addition to drug companies, physicians, other mental health personnel, advocacy groups, schools, courts and TV all have a say in determining how the written word will in fact be used and misused.[67]

Within a span of no more than 15 years, diagnoses of childhood bipolar disorder (CBD) have increase 40-fold.[68] Forty-five years ago when Frances was receiving his training in psychiatry nothing was taught about CBD: no one had experienced any cases. Now it is "the most inflated bubble in all psychiatric cases."[69] How did this happen? First, a need was created by kids who were disturbed and disturbing being encountered in clinical, school, and correctional settings. Nothing worked to correct the suffering of all concerned. A diagnosis of CBD, however, makes the false claim that it explains the misbehavior and provides a justification for medical intervention.[70]

Secondly there were those, deemed to be "false prophets" by Frances, who were influential "thought leaders" from Harvard, who "evangelized" child psychiatrists, pediatricians, family physicians, parents and teachers.[71] Then too, the drug companies that stood to benefit greatly, heavily financed their efforts. The message of these prophets was that DSM rules for diagnosing bipolar did not need to be applied to children. Instead of requiring the mood swings between mania and depression, the diagnosis could be made to include a varied group of kids who were irritable, temperamental, angry, aggressive, and/or impulsive. That resulted in labeling some children with CBD who previously would have been considered "temperamental" but normal. This opened up a whole new lucrative market for the pharmaceutical industry. Mood swings are rare in children but irritability is much more common. As one would expect, drug companies have pushed the expanded definition of CBD. Less predictable and less understandable is the enthusiastic reception of these developments by physicians, therapists, parents, teachers, advocacy groups, the media and the internet.[72]

---

[66] Ibid, 74–75.
[67] Ibid, 139.
[68] Ibid, 104. Drawing upon a different source on page 144, Frances says the 40-fold increase in diagnoses of CBD took only a decade.
[69] Ibid, 144.
[70] Ibid, 145.
[71] Ibid.
[72] Ibid.

With the more rigorous DSM-IV definition abandoned, "mood-stabilizing and antipsychotic medications were given out wildly to treat fake CBD."[73] Many harmful consequences have come about as a result of such practices. Weight gains can be rapid and substantial, averaging twelve pounds in twelve weeks. That increases the risk for diabetes and a possible reduction of one's life span. Worst of all, 2- and 3-year-old kids have been loaded up with medication to treat an extremely premature diagnosis of CBD; some have been killed by overdoses that proved to be lethal. Frances considers this to be sheer malpractice.[74]

There is also a stigma attached to being diagnosed with CBD for it implies that the child will have a lifelong illness that requires lifelong treatment. Such a diagnosis can undermine hope of achieving certain ambitions and can also reduce an individual's sense of being able to control and take responsibility for undesirable behavior. Frances concludes his account of the harms that are caused by departing from the classic definition of bipolar disorder and using it falsely to treat children, even infants, as follows:

> The CBD fad is the most shameful episode in my forty-five years of observing psychiatry. The widespread use of dangerous medicine to treat a false diagnosis constitutes a vast public health experiment with no informed consent.[75]

Frances also finds overdiagnosis of bipolar disorder in the case of adults. In this instance of doubling diagnoses in the past 15 years, the jump stems from misleading advertisements from drug companies that sell bipolar disease based on any sign of irritability, agitation, temper or elevated mood. The warnings of how perilous it is to miss diagnosing bipolar were everywhere in various media and publications addressing psychiatrists, primary care doctors, other mental health personnel, patients and families. All of this overuse of mood-stabilizing drugs has serious side effects: dangerous weight gain, diabetes, and heart disease.[76]

Clearly in the area of diagnosing and treating bipolar disorder, a great deal of harm and wasted expenditures could be avoided by practicing scientifically and morally justifiable psychiatry and medicine.

5.1.4 Autism

With a 20-fold increase in a matter of no more than 20 years, and spurred by a change in diagnostic practices, not by a sudden rise in the number of kids who are autistic, autism joins the ranks of false epidemics presently occurring.[77] Frances

---

[73] Ibid, 146.
[74] Ibid.
[75] Ibid.
[76] Ibid.
[77] Ibid, 147. At page 104, Frances, drawing upon a different source states that the 20-fold increase in the diagnosis of autism came about in 15 rather than 20 years.

identifies three causes for the epidemic of autism: improvements in looking for and identifying autism by doctors, teachers, families, and the patients; the introduction of Asperger's disorder in DSM-IV, a new category that widened the concept of autism; and for about half of the epidemic, incorrectly diagnosing children because it entitles them to receive more attention in the school and more intense mental health treatment.[78]

Actually, autism is rare. Its symptoms can be easily identified because they include the very serious symptoms of being unable to communicate and a low IQ. Those with Asperger's are strange, behaving in unusual ways, exhibiting stereotypical interests, and experiencing interpersonal problems. Since many normal people can be eccentric and socially awkward, it is difficult to distinguish them from those who could properly meet the intended characteristics of Asperger's. Frances and his colleagues estimated Asperger's to be about three times more prevalent than "classic autism" but the numbers have gone greatly beyond what was anticipated because many exhibiting variability that is normal are mislabeled as autistic, especially when diagnosed in primary care, in school settings, and by parents and patients.[79] In the various media outlets, autism was treated sympathetically. Many successful people recognized themselves being described by the criteria defining Asperger's did so proudly. Asperger's became a diagnosis that was used to explain a variety of individual differences. However, if the criteria were applied carefully, about half of the kids now being diagnoses as having Asperger's should be regarded as overdiagnosed, and if granted a repeat evaluation would have outgrown the characteristics that led to the diagnosis.[80]

Frances does see some benefits from being correctly diagnosed. Those diagnosed can receive "improved school and therapeutic services, diminished stigma, increased family understanding, reduced sense of isolation, and internet support."[81] At the same time, however, mislabeled individuals are harmed. They suffer the costs of being stigmatized and of reduced self and family expectations. It is also costly for society to have scarce and highly valued resources misallocated. As far as Frances is concerned, "school services should be based on school need, not psychiatric diagnosis."[82]

And so we see, that the overdiagnosis of autism, depression, ADHD, and bipolar disorder are all examples that illustrate the enormous financial savings and avoidance of harm that scientifically and morally sound medical practices can yield.

---

[78] Ibid.
[79] Ibid, 148.
[80] Ibid, 148–149.
[81] Ibid, 149.
[82] Ibid.

5.1.5 Social Anxiety Disorder (SAD)

It is a rare individual whose social anxiety is so totally incapacitating that anyone would define it as a mental disorder.[83] Nevertheless, ordinary shyness has been turned into one of the most common and commonly treated mental disorders, in fact the third most common with rates of 7–13%, according to how loosely it is defined.[84] Shyness is a normal human trait. If you define it as SAD, you will greatly increase the number of individuals who will be diagnosed and treated as mentally ill. That is precisely what the manufacturers of the new antidepressants did when they began to advertise their products.[85]

According to an advertisement for Zoloft, there are more than 16 million Americans who have this condition. It describes experiences all of us may have had or have sometimes: "sufferers feel anxious about meeting new people, talking to their bosses, speaking before large crowds, or drawing attention to themselves."[86] On their website, Pfizer, the maker of Zoloft, indicates that their drug can treat these symptoms. Indeed, this "disease" has been largely created, certainly its extensiveness, by the drug companies. Paxil, a drug first used for depression, was approved for use for SAD in 1999. Here is what Paxil's product director had to say about his product and SAD: "every marketer's dream is to find an unidentified or unknown market and develop it. That's what we were able to do with social anxiety disorder."[87]

Pfizer did sponsor a careful study of people suffering from SAD that tested what their drug Zoloft could do for them.[88] The study randomly assigned and divided people into four groups: two of the groups were given Zoloft for 24 weeks and two were given placebos for the same period. One of the groups taking Zoloft also were set up to receive eight 15-minute sessions to talk to a primary care physician about their symptoms. These same patients were assigned homework to undertake between sessions for the purpose of learning how to identify and overcome their social habits and fears. Similarly, one of the groups taking placebos underwent the same interventions, while the other received no counseling. For a total of 52 weeks the patients' symptoms were monitored the first 24 weeks while being treated, and then for 28 weeks after the treatments were completed.

All four groups improved significantly while being treated for 24 weeks. After that, however, those individuals who had been given Zoloft, with or without counseling, had their symptoms worsen somewhat. That result contrasted with the significant and continuing improvement of the group that did not take Zoloft but did receive counseling sessions. In assessing these data, Abramson views the patients that received Zoloft as less motivated to seek to change the kinds of reactions and

---

[83] Ibid, 152. See also Abramson, Overdosed America, 232.
[84] Frances, Saving Normal, 152.
[85] Ibid, 153.
[86] Abramson, Overdosed America, 232.
[87] Conrad, The Medicalization of Society, 18.
[88] Abramson, Overdosed America, 232–233.

interactions that gave rise to their symptoms than those who were not relying on medication. Indeed, the latter proved that they could succeed in making the necessary changes in their behavior. Abramson concludes with this rather trenchant observation:

> The discomfort of social anxiety is real, but approaching these symptoms as a fundamentally biomedical disorder and treating social skills or habits with a drug makes about as much sense (for all but the most severe cases) as treating a splinter with a narcotic pain killer instead of removing it.[89]

Indeed, Abramson goes on to say that,

> "some social anxiety disorder could be thought of not as a medical disease but as a consequence of dysfunctional patterns of social interaction shown to be amenable to significant improvement by eight 15-minute sessions of counseling with a family doctor."[90]

The consequences of having so many individuals identified as having SAD are by no means a trivial matter. The harms to individuals are considerable. After all, the very loose grounds on which people are diagnosed has been largely due to the influences exerted upon physicians and patients by the way drug companies define SAD in their ads for antidepressants. Antidepressants have very serious side effects. We remind the reader that these include committing suicide. Consider that, for every thousand persons taking antidepressants, 4.6 more committed suicide than those given placebos. Consider also that individuals who are not really sick are subject to the placebo effect and are led to believe that they need the antidepressants they are prescribed. That makes them reluctant to stop taking them, becoming therefore loyal customers set up for unnecessary complications.[91] Yet the FDA approved and approves of antidepressants for treating SAD under these circumstances.

Once again we encounter an area of medical practices that waste large amounts of money, and for the many overdiagnosed with SAD, causes more harm than benefits. All of this happens in spite of being able to help those who are anxious and not incapacitated with some 15-minute sessions with a doctor.

5.2 Medicalizing Normality: Newly Created Categories of Mental Illness

So far in this chapter we have been illustrating legitimate categories of mental illness that are being quite considerably overdiagnosed and overtreated. Now we are calling attention to new categories of mental illness that appear in the latest version of psychiatric standard setting in DMS-5. Frances refers to DSM-5 as an unhappy event for psychiatry. He expects that it will turn "diagnostic inflation into

---

[89] Ibid, 233.
[90] Ibid.
[91] Frances, Saving Normal, 156–157.

hyperinflation—further cheapening the currency of psychiatric diagnosis and unleashing a wave of new false epidemics."[92]

What the DSM-5 does is label as mental disorders common, normal behaviors such as temper tantrums, forgetfulness as one ages, gluttony, and mourning. One would think it is obvious that these are behaviors that no one would regard as abnormal or deviant. One would think also that psychiatrists would realize that there are no drugs for treating them that are helpful, appropriate, and free of harmful side effects. Nevertheless, Frances feels compelled to spell this out for his readers since there are psychiatrists that need convincing, and there are potential patients who need to be warned against being taken in by the marketing of these "illnesses." The prospect of a great deal of wasted money and serious harms should surely be avoided. Therefore, we will briefly introduce some scientific and moral grounds for doing just that.

5.2.1 Temper Tantrums

In the DSM-5, the mental disorder associated with temper tantrums is called disruptive mood dysregulation disorder (DMDD). Frances rejects the whole idea of considering temper tantrums a mental disorder.[93] For one thing, DMDD will most likely be extremely overinclusive. The children it will be used to described are children who need no diagnosis or else need a more specific one. Children have limited ways to respond to their world and temper tantrums serve quite frequently to express anger and distress. Most of the time this represents a stage in their development, or a temperamental variant, or a reaction to stress and none of these are indicative of a mental disorder. It is normally best to ignore temper tantrums. If they are severe and persistent, an evaluation seeking their cause may be required. However, "temper tantrums by themselves should never be given the status of a separate official diagnosis."[94]

Frances offers a number of reasons for viewing DMDD as a recipe for considerably increasing inappropriate prescriptions for antipsychotics. To begin with, there is very little research on DMDD. Consequently, nothing is known about: its prevalence; whether it is distinguishable from normal temper tantrums; how it relates to the multitude of other disorders exhibiting angry outbursts; its course; the preferred treatment; and how the treatment response compares with any adverse complications.[95]

Furthermore, the criteria for diagnosing DMDD are a sheer invention and not restrictive enough. They will lead to misdiagnosing normal children who are experiencing a state in their development and who are normally temperamental: "there is

---

[92] Ibid, 170.
[93] Ibid 177.
[94] Ibid.
[95] Ibid, 178.

no bright line distinguishing normal temper tantrums from abnormal ones."[96] Also what is regarded as appropriate varies quite markedly for different families, subcultures, and with respect to developmental periods.

Finally, how the diagnosis of DMDD will be made will vary a great deal in accord with what clinicians, families, schools, and peer groups can tolerate. In the situations in which temper tantrums occur, it will be easy to overlook that most children "outgrow their developmental or situational temper problems and gradually acquire self-control and better ways of getting their needs met."[97]

Frances allows for the possibility that in extreme circumstances, or rarely, to reduce explosive temper outbursts, antipsychotic drugs may be called for. However, for the many normal children who will have "developmental or situational storms or are irritable for other reasons... antipsychotic drugs could turn out to be the most dangerous epidemic caused by DMS-5."[98] As we learned in our discussion of child bipolar disorder, antipsychotics have serious side effects, including the risk of fatal doses.

In the end, then, DMDD as a diagnosis lacks scientific credibility, and the harms of antipsychotics to treat it, cannot therefore be defended on moral grounds. Again, there is a financial cost that can and should be avoided.

5.2.2 The Forgetting of Normal Aging

Frances begins an account of his objections to making a disease out of becoming more forgetful as one ages with a humorous description of the kinds of things he and his wife have a tendency to be forgetful about, such as misplaced items and where they parked their car. Yet at the same time, they retain their important mental capacities and make use of them to engage in the many activities that characterize their busy and eventful lives. There are also many physical activities that decline as one ages such as climbing stairs, pushups, and sleeping soundly, yet the loss of physical prowess and strength is taken to be inevitable, not a sickness. In contrast, the DMS-5 has made a mental disorder out of some loss of one's cognitive abilities: the name for it is mild neurocognitive disorder (MNCD).[99]

The intention in creating the disorder of MNCD is to identify individuals exhibiting some signs of mental decline who don't have dementia yet but are at risk of developing it: the idea is to prevent dementia. That is something Frances could heartily endorse if a treatment for accomplishing this existed, and if there were some effective method for predicting one's future cognitive powers. As Frances observes: "without a laboratory test, the diagnosis of MNCD will be wildly inaccurate, pulling in many people who are not headed for dementia." And he adds, since there are no effective treatments at the present time, to be told that "you are (only possibly) at

---

[96] Ibid.
[97] Ibid.
[98] Ibid, 178–179.
[99] Ibid, 179–180.

risk for later developing Alzheimer's would provide little or no benefit—but would create needless worry, testing, treatment, expense, stigma, and insurance and disability issues."[100] What is more, the drugs for Alzheimer's have been profitable for drug companies but of little or no efficacy for those receiving them."[101]

The experts who put together DSM-5 thought that MNCD would benefit their patients. But the corporations marketing drugs and diagnostic tests will develop a plethora of tests and medical treatments that will likely be useless, at least early on. Thus it is these corporations that stand to benefit from the new MNCD diagnosis rather than patients and taxpayers.[102]

MNCD is an instance in which much money can be saved by completely avoiding the harms this scientifically unwarranted and non-beneficial diagnosis: justice demands no less.

5.2.3 Gluttony

DSM-5 has created a so-called binge eating disorder (BED). This diagnosis has a very low threshold. Individuals qualify for this alleged mental illness by indulging in one binge a week and that for as little as three months.[103] By these criteria, Frances says he has had BED since he was a teenager. He is certain he is not alone. The diagnosis of BED may add "20 million fake mental patients in the U.S. alone."[104] That will happen as a result of the drug-company-sponsored "education" for doctors and the public, designating gluttony as an illness. Frances notes that gluttony has been regarded as a sin. We should remind ourselves that making gluttony an illness does not hold us responsible for our behavior. Frances does act responsibly by keeping his weight down by exercising several hours a day and skipping some meals when necessary.

BED is a psychiatric attempt to address the serious threat to health posed by obesity. Frances asserts, however, that psychiatry will not help to curb obesity by addressing it with what Frances calls phony psychiatric labels: BED has the familiar three-strikes-you're out combination of inaccurate diagnosis, no effective treatment, and drug side effects: "public policy should aim at changing notions about what to eat and about exercise with the kind of campaign that has contained smoking."[105]

By inventing BED and making it a mental illness, DSM-5 has opened the door, potentially very widely, to diagnoses and treatments that will waste money and time, and can only cause harm from the side effects of drugs. Once again, we have a practice being put forward that is not just or justified from either a scientific or moral standpoint.

---

[100] Ibid, 181.
[101] Ibid.
[102] Ibid, 181–182.
[103] Ibid, 182.
[104] Ibid.
[105] Ibid, 183–184. One page 183, Frances offers a variety of specific public policy suggestions for curbing obesity.

5.2.4 Mourning

DSM-5 has medicalized grief. DSM-5 has facilitated a diagnosis of major depressive disorder (MDD) for those who are bereaved, a diagnosis that can be made in the first weeks of their loss.[106] As Frances points out, in the normal process of grieving people exhibit the same symptoms characteristic of those suffering from clinical depression: "feeling sad, losing interest, trouble sleeping and eating, reduced energy, difficulty working—this the easily recognizable, classic picture of grief."[107] The diagnosis of MDD requires that the bereaved individual begins to be suicidal or delusional, or exhibits symptoms that are serious, prolonged, and incapacitating.[108]

Medicalizing grief and falsely labeling it as a mental disorder has a number of harmful consequences. Among these is interfering with the working through of one's loss and thereby reducing one's reliance on the longstanding rituals of one's culture that provide consolation for one's grief. At the same time, this trivializes the life one has lost. Depersonalized medicalized mourning is no match for the solemnity of well-tested death rituals that assist, by far the most of us, to recover from our losses.[109]

The harmfulness of this unnecessary diagnosis of MDD is exacerbated by the recourse to antidepressants. Doing that, ironically enough, subjects one so diagnosed to the risks of suicide, among others, and does so without benefits that exceed the effect of taking a placebo.

Medicalizing grief as MDD, then, is still another instance of a financially wasteful, harmful, and unnecessary and unfortunate creation of DSM-5. Again, there are no sound scientific and moral reasons for going this way.

5.2.5 Behavioral Addictions

What DSM-5 has done with addictions is to expand their scope so broadly that any passionate interest or attachment can qualify as a mental disorder. As Frances indicates, behavioral addiction is a fundamentally flawed concept since it encompasses all of us: "repetitive pleasure seeking is part of human nature and too common to be considered a mental disorder."[110] Behavioral addiction provides a "sick role," an excuse for impulsively indulging in some pleasure or pleasures. It allows one to claim that one is addicted and cannot help behaving as one does. In short, it undermines taking responsibility for one's actions.

---

[106] Ibid, 186.
[107] Ibid, 186–187.
[108] Ibid, 187.
[109] Ibid.
[110] Ibid, 189.

Frances sees no way to distinguish compulsive non-pleasurable repetition from impulsive self-indulgence.[111] The predictable widespread, misapplication of the category of behavioral addiction greatly exceeds any possible benefits from its application. Frances puts his finger on a very significant harm of labeling people as behavioral addicts:

> A vibrant society depends on having responsible citizens who feel in control of themselves and own up to the consequences of their actions—not an army of 'behavioral addicts' who need therapy in order to learn to do the right thing.[112]

If psychiatrists and physicians will recognize the pitfalls of labeling people as behavioral addicts for their repetitive pursuits of various ordinary, pleasurable activities, whether a sport, a hobby, a harmless form of recreation, etc., a great deal of money and time can be saved to meet genuine medical needs, including those that arise from the truly destructive addictions to alcohol and drugs.

5.3 Where We Are and Where We're Headed

This chapter should leave us with little doubt that there are a number of mental illnesses that are grossly overdiagnosed and overmedicated. Curbing these practices would save a lot of money, and save many normal individuals from the harms of drug side effects, psychological distress, and diminished independence and accountability. As also in chapter four on physical illnesses, we have only introduced the reader to a sample of interventions that involve large numbers of patients and are particularly harmful, scientifically unjustifiable and very costly We also chose a sample of recently minted categories of mental illnesses, all of which have the potential to be extremely harmful, financially wasteful, and turn many normal people into patients treated for being mentally ill.

Reducing the extensive amount of overdiagnosing and overtreating individuals, and undue medicalization of what should be viewed as physically and psychologically normal conditions, behaviors, and traits, will not be easily brought about. There are powerful forces at work that influence the standards and medical information so many physicians presently conscientiously follow. There are significant changes that will have to occur in the policies and behavior of drug and medical device companies, the FDA, hospitals, and the federal government. The sheer understanding of justice as a demand for mercy will also have to be cultivated by all who are engaged in providing medical care, including those named above.

In the next chapter, we will examine what policies and behavior will have to be curbed, and in some cases stopped entirely, if overdiagnosing, overtreating and over medicalizing, and the harms and costs they bring about, are to be replaced by policies and behaviors that are just, that is to say, morally and scientifically justifiable.

---

[111] Ibid, 190.
[112] Ibid, 191.

Chapter 6
Practices and Policies in the U.S. Health Care System that Are Scientifically and Ethically Unjustifiable: They Should Not and Cannot Persist

Abstract This chapter continues identification and analysis of medical practices and policies that are very costly and should not be allowed to persist because they are unethical, either because they are unjustified on scientific grounds or because of violation of justice and hence, unethical. Such activities and policies in the U.S. healthcare system contribute enormously to over diagnosis, over medication, and mistaken diagnoses and treatments. The evidence for what brings all of this about is abundantly documented in this chapter.

This chapter is absolutely essential. Though physicians largely make the decisions that lead to overdiagnosis, unnecessary treatments, and overmedicalization, and though there are individual physicians and those in the Choosing Wisely Movement engaged in curbing these practices, doing so sufficiently requires changes from other participants in the U.S. health care system. Why this is so, and who those other agents and organizations are, will occupy us in this chapter. The physician Norton Hadler characterizes these powerful forces as follows:

> The American health-care system is as entrenched and wealthy as it is perverse. It is also unsustainable. It is hell-bent for collapse and its leadership is much more committed to and adept at self-service than it is foresighted or ethical. The result will be turbulent, and many a lovely servant of the ill, and many an ill patient will be sorely served during this time.[1]

I agree with Hadler that the unjustifiable activities and policies that make unsustainability a reality are at the bottom unethical, very much lacking in mercy which alone can make what is unjust just. Alas, many morally conscientious physicians and patients are presently being "sorely served" as Chaps. 4 and 5 plainly illustrate. As this chapter will document, the influences upon physicians and patients that bring about unjustifiable diagnoses, treatments and medicalization, are hard to detect. These potent influences emanate from the manufacturers of drugs, medical devices and equipment, hospitals, the Food and Drug Administration (FDA), the way malpractice is handled, and the involvement of the federal government. We

---

[1] Norton Hadler, The Citizen Patient: Reforming Health Care for the Sake of the Patient, Not the System (Chapel Hill, NC: The University of North Carolina Press, 2013), 222.

begin examining these potent influences by considering those of the pharmaceutical industry.

6.1 Documentation of Undue and Unethical Influences and Practices of Drug Companies and Their Impact

As we have gladly acknowledged previously, drug manufacturers have been, and continue to be, sources of curative, life-saving, and pain-relieving medications, and of vaccines that prevent and reduce diseases. Our focus is on practices and policies that cannot be scientifically, legally, and ethically justified. Why this focus? Because unfortunately and tragically what is at stake is the existence of far too many medications that are simply non-beneficial, and given their side effects, too often harmful, sometimes seriously, even fatal. Ironically, if these breaches of ethics, law and science are not halted or greatly reduced, they will be financially unsustainable and the excessive pursuit of profits will prove unprofitable.

The psychiatrist Allen Frances takes a dim view of the consequences of allowing drug companies free reign to pursue profits as their primary goal rather than the public's well being: "Government, doctors, patients, the media, advocacy groups—all were largely bought off by the drug company money and power."[2] The failure to be aware of the untoward influence of what he refers to as "commercialized care" prompted the physician and professor emeritus of Harvard Medical School, Arnold Relman to write a book suggesting ways to reform the U.S. health care system. His chapter on "The Consequences of Commercialized Care" begins with this quotation: "for the love of money is the root of all evil" found in I Timothy 6:10 (Kings James Version).[3] The New Revised Standard Version of this biblical passage is more appropriately nuanced: "For the love of money is a root of all kinds of evil."

These are strong words with which to depict the unjustifiable, unethical activities discussed in this chapter. However, the persistence of these activities threaten the very viability of the entire healthcare system. Our portrayal of why this is so begins with how drug companies carry on the research they sponsor.

---

[2] Allen Frances, Saving Normal: An Insider's Revolt Against Out-of-Control Psychiatric Diagnosis, DSM-5, Big Pharma and the Medicalization of Ordinary Life (New York: HarperCollins Publishers, 2013), 106.

[3] Arnold S. Relman, A Second Opinion: Rescuing America's Health Care; A Pleas for Universal Coverage Serving Patients Over Profit (Cambridge, MA: Perseus Books Group, 2007), 41.

6.1.1 Rigged Medical Research

Like an increasing number of physicians, John Abramson sees the American health care system moving toward its breaking point. Among other factors contributing to this, Abramson singles out as the most important one the following one: "the transformation of medical knowledge from a public good, measured by its potential to improve our health into a [commodity] measured by its commercial value." Abramson attributes this transformation to the "commercial takeover of the process by which 'scientific evidence' is produced."[4] What I wish to stress that this process of gathering what is to count as medical knowledge subverts what should be reliable scientific findings by employing <u>unethical methods</u> of collecting, disseminating, and interpreting data. We being with the ways studies are rigged.

6.1.1.1 Using Ideal Patients

Ideal patients are individuals who are [more] likely than others to get better and therefore exaggerate how beneficial the drugs being tested are. That means also that the applicability of these drugs to the various patients commonly seen in medical clinics is not actually known. Furthermore, the relation between benefits and risks may markedly change when we move from one population to another. Consider for example, that [anti-arhythmia] drugs were shown to be effective at prolonging life in patients with severely abnormal heart rhythms. But when they were more widely prescribed for patients with mildly abnormal heart rhythms after they had heart attacks, trials on these patients discovered that—"to everyone's horror"—the use of these drugs resulted in an increased risk of dying.[5] Using patients selectively to favor a positive outcome can also distort the cost/benefit ratio of prescribing a particular drug studied in this manner.[6] In short, clinical trials selecting patients who are unrepresentative yield ungeneralizable results that can be of little or no benefit, and/or very harmful.

6.1.1.2 Testing Drugs Against Something Relatively Ineffective or Rendered Ineffective

As we have noted in Chaps. 4 and 5, drug companies tend to test their drugs over against placebos (sugar pills) that contain no medicine. This provides a clear opportunity to find a drug at be at least somewhat more effective. Whenever the condition

---

[4] John Ambramson, <u>Overdosed America: The Broken Promise of American Medicine</u> (New York: HarperCollins, 2004), 91. See Chap. 7, pages 93–110 for an extensive discussion of the ill effects of commercialization of medical knowledge.

[5] Ben Goldacre, <u>Bad Pharma: How Drug Companies Mislead Doctors and Harm Patients</u> (London, Great Britain: HarperCollins Publishers, 2012), 176–177.

[6] Ibid, 178–179.

being studied elicits a considerably beneficial placebo effect, the drug being tested will benefit from that same placebo effect. What is also quite frequently carried out are comparisons of a new drug with one that is known to be virtually useless or with one that works fairly well but employ it at an absurdly low or absurdly high dose so as to render it either quite ineffective or fraught with undesirable or harmful side effects.[7] Obviously all of these [ploys] favor the drugs being tested.

6.1.1.3 Conducting Trials that Are too Short

The short term effects of drugs can be quite different from the long term effects. Brief trials can distort the benefits of drugs as well as their harmfulness. The weight-loss drug Fenphen caused weight-loss during positive brief trials. When observed for longer periods, they caused those treated to develop heart valve defects. Similarly, the drug Valium will alleviate anxiety for a short period, but in a period of months or years that follow the short-term benefits decrease and patients tend to become addicted.[8]

6.1.1.4 Stopping Trials Early or only Publishing Short Versions of a Lengthier Trial

These are still more strategies to increase the likelihood of obtaining and presenting positive outcomes for a drug being tested. For example, a trial comparing a new pain-killer called Celecoxib with two older drugs for relieving pain over a period of 6 months revealed that it led to fewer gastrointestinal complications. As a result, more doctors prescribed it. However, a year later, it was discovered that the original trial was found to be of no benefit in what was a year-long trial. Only the favorable results of the first 6 months of the trial were published.[9]

Though it is sometimes prudent or necessary to stop a trial, one has to have good reasons for doing so. Consider a trial of using the drug Biosoprolol during blood-vessel surgery. It was stopped early because only two patients on the drug experienced a major cardiac event as contrasted with eighteen of those taking placebos. However, the benefit had been overstated. Two [years] later, larger studies found that the drug was not at all beneficial. Early stoppage of the initial trial had yielded incorrect results.[10]

Goldacre cites these systematic reviews of shortened trials: one published in 2010; another in 2005; and another in 2008.[11] The review published in 2010 found that trials that were stopped early overstated the benefits being tested by a quarter as

---

[7] Ibid, 180–182.
[8] Ibid, 182–183.
[9] Ibid, 184.
[10] Ibid, 185.
[11] Ibid, 186.

compared with trials that were not stopped early. The review published in 2005 discovered that trials halted early have doubled since 1990. The review published in 2008 brought to light that, of all the randomized tests of treatments for cancer published in the previous 11 years, more than half of those halted early were published in the past 3 years. Furthermore, 86% of these trials shortened early were carried out to bring a new drug to market as quickly as possible. Cancer drugs can be very quickly profitable. These reviews reveal how seriously and extensively medical research is rendered misleading by these practices.

6.1.1.5 Trials Testing Multiple Outcomes that Allow for Switching the Main Outcome to One that Yields a Positive Result

By measuring, say, a dozen outcomes, you have a dozen chances of obtaining at least one positive result. Under these circumstances, you have not designed a study that is scientifically valid. To be able to decide that an outcome is statistically significant requires a study that measures one outcome. A study that measures multiple outcomes is biased by design because it increases the chances of finding more positive results than there really are.[12]

In 2009, a group of researchers investigated the drug Gabapentin to see if the trials reported as primary outcomes had been pre-specified as such.[13] To begin with, they found that about half of the trials had not been published. Of the twelve trials that were published, only eleven of twenty-one primary outcomes pre-specified as such appeared in the published articles: six were not reported at all; four were reported as secondary outcomes.

This deceptive and scientifically invalid practice is alarmingly widespread. In 2004, a study was published of trials approved by ethics committees of two cities over 2 years in all areas of medicine.[14] This research found that about half of the outcomes were not correctly reported. Among the published trials, two thirds of them revealed that at least one pre-specified primary outcome had been switched. This was done deliberately: positive outcomes were more than twice as likely to be truthfully reported.

That such activities take place at all, never mind so frequently, is no small matter: they can and do harm patients and are financially costly.

Consider, for example, the antidepressant Paroxetine. In 2008, a group of researchers studied one trial of Paroxetine that had been published, touting positive results.[15] Systematic reviews up to 2008 backed that published outcome. All the while the drug was being prescribed for several years. The claims on behalf of the drug's efficacy proved untrue. There were two primary outcomes and six secondary ones specified as such in the original protocols. When the trial ended, none of these

---

[12] Ibid, 200–202.
[13] Ibid, 202–203
[14] Ibid.
[15] Ibid, 204.

outcomes yielded a difference between Paroxetine and placebo. The trial measured nineteen additional outcomes of which only four were positive outcomes: these were reported as being the main outcomes, though they clearly were not.

Now what resulted from the use of this useless drug? In the United Kingdom, the regulative authorities advised doctors not to prescribe Paroxetine to patients under eighteen because it increases the rate of suicide in this population.[16] Furthermore because of false claims of safety for a drug they know to be unsafe, GSK was fined 3 billion dollars by the FDA. Yet that large amount of money is not that much shen it is compared with the 12 billion dollars brought in by GSK's sales of Paroxetine.[17] Recall that in Chap. 5 we noted that antidepressants of the same type are among the drugs being grossly overprescribed for minors and adults. This happens because depressions are being markedly overdiagnosed, and the antidepressants continue to be aggressively and misleadingly marketed.

6.1.1.6 Deceptive Bundling

Outcomes can be bundled together to create one composite outcome. That can lead to reporting misleadingly heightened benefits or misleading diminution of harms for the drugs that were investigated. For example, there was a trial called [UKPDS] to see if intensively managed blood-sugar levels would benefit diabetes patients.[18] The study found no benefit for preventing death and diabetes-related health. However, a composite outcome also was reported that touted a 12% reduction in untoward events achieved because of reductions in events of much less significance than death and diabetes-related deaths.

Yet, the treatment given in the trial came to be widely regarded as beneficial by physicians. How come? Consider how it was reviewed. A study of thirty-five diabetes review articles that cited the UKPDS trial discovered that twenty-eight reviews took note of the benefit reported for the composite outcome; only one mentioned that mostly trivial outcomes accounted for that; only six found that there was no benefit for death, surely the most important outcome.[19]

6.1.1.7 Analyses that Ignore Dropouts

This is yet another method employed to exaggerate the benefits of a drug. To test whether a given drug is one that physicians should be prescribing for a specific condition, you should find out all the effects taking that drug will have on those who

---

[16] Ibid, 60–61. As discussed in Chap. 5, this whole class of antidepressants have a heightened risk of suicide for adults also.

[17] Ibid, 346. For a detailed account of how and how long GSK misled authorities, doctors and patients about the safety of Paroxetine, see Ibid, 58–63.

[18] Ibid, 196–197.

[19] Ibid, 197–198.

receive it. After all, those issued the drug you are studying may drop out because of the way in which the drug affects them. People could be dropping out for various significant reasons: they may believe the drugs are not working' or they may have been experiencing truly harmful side effects; or they may have died from taking the drug being tested, failure to include these data will overstate any beneficial effects and/or understate any harmful effects a treatment may yield.

An analysis that includes dropouts is called "intention to treat;" an analysis that ignores them is called "per protocol" analysis. There was a systematic review of the trial reports of companies submitted to the Swedish drug regulators and those that were subsequently published.[20] All but one of the submissions to the regulators had done both "intention to treat" and "pro protocol" analyses. However, of the academic publications, all but two reported only the "per protocol" analysis, thus overstating the benefits. These articles are what the physicians read and use to guide their decisions. This is another perversion of medical knowledge. Academic journals should not publish such deceptive articles.

6.1.1.8 Reporting Misleading Results by Means of Analyses that Create Subgroups that by Chance had a Desired Result

The trials carried out to test the effectiveness and/or safety of a particular drug may not yield the outcome that was sought. Yet, by breaking up the data in subgroups of patients, a group of subjects may by chance have a result desired by those conducting the research. Over time, quite a number of such analyses have deleteriously mislead physicians whose prescribing has been "informed" by such analyses.[21]

6.1.2 Deceptively Reporting Research Results

There are trials in which a drug has a negative result in that its primary outcome shows no benefit: these negative results may not be reported at all. In 2009, researchers studied seventy-two trials submitted in 1 month that yielded no benefit for their primary outcome.[22] What they found was that twenty-eight of the seventy-two published trials provided no numerical results for the main outcome whatsoever; only sixteen reported the negative results anywhere; and only nine did so in the abstract of the article. It is imperative to realize that many busy physicians may often only read the abstract to be informed of new drugs they should consider prescribing for their patients. Being misinformed like this may prove harmful to their patients, and certainly of no more benefit than a placebo effect which may be quite minimal or of short duration.

---

[20] Ibid, 199.
[21] Ibid, 205–210. See also, Ibid, 210–212, for unreliable subgroupings of trials rather than patients.
[22] Ibid, 220–221.

Even more disconcerting is to have this same kind of spin in systematic reviews and metanalyses that are regarded as the best source of reliable evidence. That was revealed in a study comparing industry-funded reviews with those of the independently funded Cochrane Collaboration.[23] The industry-funded reviews recommended a treatment in question unreservedly; in no instance did the Cochrane reviews do so. This difference showed up in the conclusion section of the reviews: there the industry-funded reviews offered a narrative that favored the drugs that were tested. These reviews also tended to give little or no attention to methodological shortcomings, and the possible bias of the studies being reviewed, whereas the Cochrane reviews paid much more attention to these matters. As Goldacre rightly observed, the "biases associated with industry funding penetrate very deeply into the world of academia."[24]

Deceptive depictions of industry funded research published in medical journals also is achieved by failing to report data that portray an unfavorable characteristic of the drug being advocated. How harmful it can be to prescribe drugs so falsely put forward for use is what the example described below dramatically illustrates.

The drug Vioxx, a product of pharmaceutical company Merck, was approved by the FDA for relieving arthritic pain. Serving as evidence for commending its use, the Vioxx Gastrointestinal Outcome Research Study (Vigor) was published in 2005.[25] The purpose of this research was to find out if Vioxx had fewer serious gastrointestinal complications than the over-the-counter painkiller naproxen (Aleve). What was published deliberately left out some very critical data: gastrointestinal problems were reported for the full duration of the study; cardiovascular problems were not reported fully, only those issuing from an early cutoff date for reporting those particular results. The article did not clearly indicate that these very important results were missing, Readers, therefore, were given a misleading picture of the drug's benefits and risks.[26]

The FDA had this missing data in its possession and did not reveal that it did. Analyzing all of the data generated by the Vigor study, an FDA statistician found that those who took Vioxx experienced more than twice the cardiovascular complications than those who took Aleve. The likelihood that this difference was due to chance was 1 in 10,000 (p.=0001).[27] The article in 2000 on the Vigor study concentrated on heart attacks and did not report on the full range of cardiovascular problems experienced by those in the study, a serious omission. These cardiovascular complications are exceedingly harmful indeed: besides heart attacks, they include sudden cardiac death, stroke, unstable angina, transient ischemic attack, arterial blood clot, and venous blood clot. Having observed the greater risk of these side effects incurred by Vioxx as compared with Aleve, an FDA reviewer concluded that

---

[23] Ibid, 222.
[24] Ibid.
[25] Abramson, Overdosed America, 35.
[26] Goldacre, Bad Pharma, 95.
[27] Abramson, Overdosed America, 35.

Aleve could be the drug preferred for us, not Vioxx.[28] What the complete data of the Vigor study showed, that for every hundred people who had a history of cardiovascular disease who received Vioxx rather than Aleve, between seven and eleven additional serious cardiovascular complications would occur each year.[29]

Doctors reading the article could have no knowledge of the magnitude of these very harmful risks of being treated with Vioxx. Missing data, therefore, can prove fatal, painful, and debilitating for a great many patients. Merck, not the FDA, withdrew Vioxx on September 30, 2004. A considerable fine for deceptive advertising had the effect of making public the missing data and much else. Then in November, 2004, Congress heard the following testimony from Dr. Graham, the Associate Director for Science at the FDA's Office of Drug Safety: Vioxx was responsible for 88,000–139,000 excess cases of heart attack and stroke; and intentionally withheld information it had that documented the harmful and lethal side effects of Vioxx.[30]

Vioxx is but one example of the enormously harmful consequences of misleading publications. Goldacre provides us with some especially egregious examples, including one that caused more than 100,000 deaths.[31] What deeply troubles Goldacre and should trouble all of us, particularly those who intentionally suppress research with negative outcomes, is that,

> 'publication bias'—the process whereby negative results go unpublished—is endemic; and that regulators have failed to do anything about it, despite decades of data showing the size of the problem.[32]

Goldacre supports his claim about ample data by citing and summarizing the results of some of the abundant scientific studies that document the extensiveness of missing data and publication bias.[33] Goldacre also discusses how difficult it is to obtain missing data from drug companies and the very untoward consequences of prolonged periods of not knowing what is embedded in that missing data.[34]

Goldacre maintains that the problem of this common and widespread research misconduct is "universally recognized but nobody has bothered to fix it."[35] This is true for regulators, the FDA among them. They possess unpublished trials and yet they obstruct those who seek information beyond what has been published. It is also true that drug companies withhold trial results from the regulators, and thus undermine their ability to carry out their mandate to ensure that only safe, efficacious drugs are approved. Additionally, university administrators permit contracts that

---

[28] Ibid, 34–35.
[29] Ibid, 35.
[30] William Faloon, Pharmocracy: How Corrupt Deals and Misguided Medical Regulations Are Bankrupting America—and What to Do About It (Mount Jackson, VA: Practikos Books, 2011), 216–217. Faloon here speaks of tens of thousands of deaths caused by Vioxx.
[31] Goldacre, Bad Pharma, 5–11.
[32] Ibid, 7.
[33] Ibid, 19–29.
[34] Ibid, 81–95.
[35] Ibid, 80.

explicitly indicate that the sponsor can control the data. Registering all trials before they are conducted has been recognized as a check on missing data but it has not been enforced. Nor are academic journals abiding by the policy they profess as they continue to publish unregistered trials. In all these instances, we are confronted with very troubling and consequential ethical lapses.

6.1.3 Corrupting and Biasing Medical Education, Medical Academics and Sources of Medical Knowledge

Since doctors are making decisions for approximately 80% of all healthcare expenditures, the medical industry makes every effort to exert a financially beneficial influence on these decisions. That influence begins when doctors enter medical school and does not end until they retire. Medical students and residents are offered free lunches by drug companies that use the occasion to present what Abramson calls "subtle and not so subtle infomercials."[36]

These practices persist in the context of continuing education. Doctors are invited to free suppers and conferences, sometimes in splendid tropical settings, to be "informed" about new medical products.[37] A study of a group of doctors who were followed before and after attending a drug company's expense paid symposium in a much visited vacation site.[38] Before going on this trip the majority of doctors expressed confidence that their prescribing behavior would not be changed as a result of what transpired at this symposium. They were wrong: their prescriptions of what this company was selling increased threefold.

Drug representatives (drug reps) also visit physicians in their offices trying to persuade them that their drugs are superior to others. They come laden with free samples of their drugs and reprints of articles favoring their drugs but none that do not favor their drugs. They also give out gifts that include free meals. Drug reps come prepared to offer objections to drugs that compete with the ones they are hawking. They come as well with knowledge of what the doctors they target are prescribing.[39] Some of these drug reps are paid on the basis of their results.

Do they enjoy success? Yes, quite markedly! A systematic review of twenty-nine studies documents a high degree of success.[40] Out of the total of twenty-nine, seventeen discovered that physicians visited by drug reps will be more likely to prescribe the drug being recommended: six had mixed results; the rest found no difference. In no instance, did the studies find a drop in prescribing. Noteworthy was the finding that the physicians in contact with drug reps are more likely to incur

---

[36] Abramson, Overdosed America, 112.
[37] Ibid.
[38] Goldacre, Bad Pharma, 282.
[39] Ibid, 279–284. Here see a detailed account of drug rep activities.
[40] Ibid, 275.

higher prescribing costs and are less likely to be in line with the guidelines for best-practice prescribing. Despite what is actually happening, physicians believe that what they are prescribing is not influenced by drug reps, while allowing that other physicians may be. And, the more drug reps physicians see, the more inclined they are to believe that they have no influence on them.[41]

The deleterious and costly effects of drug reps on the practice of medical practitioners goes quite beyond increasing the use of their particular drugs. A published review of doctor-drug company interactions reveals that.[42] The more contact physicians have with drug reps, and the more they possess drug samples they have been given, the more likely they are to prescribe newer and more costly drugs, and the less likely they are to prescribe generic drugs. Furthermore, the more doctors see drug reps, the less likely they are to recognize false claims about a given drug; the more likely they tend to prescribe more drugs overall; and they also are about 15 times more likely to ask that hospital pharmacies have a supply of drugs of a specific company.

Negative effects emanating from the influence of drug companies upon the practice of medicine occur in yet another very disturbing way. Since medical practitioners cannot possibly keep up with the enormous amount of published research on all the different medical conditions, the medical profession establishes guidelines designed to specify the nature of good care. Panels of nationally acknowledged clinical experts create these standards of care.

After citing a couple of studies that reveal flaws in the work of these panels, Abramson calls our attention to the results of a study he regards as "the most damning."

> The study found that four out of five experts who participate in the formulation of clinical practice guidelines have financial relationships with drug companies, averaging more than 10 such relationships each. A whopping 56 percent of the experts "had relationships with companies whose drugs were considered in the guidelines they authored."[43]

Imagine that! These 59% of the experts are behaving like judges carrying on a financial relationship with litigants whose cases they were hearing.

Chapter 4 documented the many grave harms inflicted upon patients by clinical standards that lead to overdiagnoses, overtreatments, overmedicalizing, and unjustifiable, unreliable, ethically compromised guidelines for what physicians do. Having such unscientific, unreliable, ethically compromised guidelines for what physicians do, certainly makes it extremely difficult for them to practice efficacious, compassionate medical care at the present time, and that is an understatement!

Advocacy for particular drugs, and the use of drugs more generally, extends to physicians in academic positions who are in financial relationships with drug companies and thus enhance the pharmaceutical industry's influence on medical instruc-

---

[41] Ibid, 274–275.
[42] Abramson, Overdosed America, 1225–126. See also Goldacre, Bad Pharma, 274–276.
[43] Abramson, Overdosed America, 127–128.

tion in medical schools and teaching hospitals.[44] Lectures sponsored by a drug manufacturer are two-and-a-half to three times more inclined to refer to the sponsor's drug in a favorable way and drugs of competitors in a neutral or negative way. Furthermore, doctors who receive honoraria from drug companies speaking and for their research are four to nine times, respectively, to endorse that manufacturer's drugs for use in their hospital.

A rather notable investment in medical education to the tune of three million dollars was awarded to Massachusetts General Hospital.[45] In return for that gift, the hospital's pain center was renamed the MGH Perdue Pharma Pain Center. The money will be used to support educational activities, including continuing education. The agreement also specifies that MGH pain specialists conduct seminars on pain control for the purpose of continuing education. To do this, these pain specialists would use a curriculum designed by Perdue, one that, among other things, serves to encourage doctors and pharmacists to prescribe the pain-killer Oxyconton, a drug Perdue happens to manufacture. It should could come as no surprise that George Annas, a medical ethicist, questioned the whole idea of MGH putting its name on a curriculum written by outside people, in particular by a drug company that can potentially run it for its own purposes.[46] Advocating Oxyconton is not to be taken lightly. Anyone following the news knows that Oxyconton is an opioid, an addictive drug, and being addicted to it has led far too many people to purchase the cheaper addictive drug heroin. Deaths from overdosing on prescribed addictive pain killers and heroin are on the rise at an alarming rate even as I write this. All the while, safer, non-addictive, and cheaper pain relievers exist.

It is also disconcerting that a study in 2011 informs us that 88% of those who attend industry supported educational activities recognize that this commercial sponsorship introduces bias, but only 15% favored banning this kind of free teaching. The majority of doctors did not wish to pay properly for this continuing education.[47]

Does it pay pharmaceutical companies to spend large amounts of money as they do for educating doctors? Yes indeed it does! Merck researched this internally and found that they got back almost two dollars in earnings from the increased prescriptions for every dollar they spent in the discussion groups doctors led.[48] The expenditures are quite considerable. In 2008, the drug industry's body accredited to offer continuing education did so to the tune of 100,000 teaching activities, totaling more

---

[44] Ibid, 122–123
[45] Ibid, 121.
[46] Ibid, 121–122. See also Goldacre, Bad Pharma, 311–321 for a thorough discussion of drug company's use of doctors for continuing education and what to do to stop it. On pages 321–340, see an extensive account of the dire consequences of drug companies financing doctors to, in effect, market their products, including what can be done about it.
[47] Goldacre, Bad Pharma, 320.
[48] Ibid, 319.

than 760,000 hours: for this more than half was directly paid for by the drug industry.[49]

With their rich purses, drug companies also morally corrupt leading physicians by luring them into putting their well recognized names on research that was entirely conducted and written for publication in a top medical journal by drug company employees. That is a practice that is called plagiarism when students try it and it is definitely not condoned. It is astonishing then, that academics would do this at all and equally astonishing that it is not uniformly banned. However, a survey in 2010 of the fifty leading medical schools in the U.S. revealed that only thirteen of them had a policy of not permitting their scholars to put their names on articles they did not prepare and write.[50]

This loss of a proper moral compass was clearly displayed by the academic who put his name on the infamous Vioxx study as the lead author. He disavowed any responsibility for the failure of the article adequately to document the deaths of patients who received the drug. His defense of his "authorship" was to point out that Merck designed, financed, carried out the trial, and wrote the article: he only edited the completed manuscript Merck submitted to him.[51] Imagine that for justifying one's actions! The act of claiming authorship for what one has not researched or written is portrayed as if it is completely and commonly acceptable, certainly nothing he should be blamed for doing. Consider that this practice of claiming authorship for what one has not at all researched or written extends even to medical textbooks produced by drug companies.[52]

What has happened to the moral consciences of those employees who make a living writing these articles and textbooks that are composed in order to enhance the sales of particular drugs, even in instances known to be harmful to those who receive them? Drug companies shouldn't be able to find or long retain anyone to perform these tasks. Such losses in moral integrity are yet another threat to the very moral fabric necessary for achieving and sustaining quality medical care, and, for that matter, sustaining the economy of the whole society. The big banks certainly did quite a job on the entire U.S. economy by their indulgence in unethical and illegal pursuit of ever more wealth.

Academic journals serving the medical community could do a great deal to curb many of the publications of unethical, scientifically unreliable research. That hasn't been happening and probably will not as long as medical journals continue to be so heavily dependent financially upon the revenue they glean from the pharmaceutical industry. They even profit from favoring drug company research for publication over government sponsored research.

Why is that so? Drug companies are eager to buy reprints of their articles from the journals that publish them. These are highly valued marketing tools that drug reps distribute to doctors; their companies have been known to pay as much, or even

---

[49] Ibid, 317.
[50] Ibid, 299.
[51] Ibid, 296.
[52] Ibid, 296–298.

more, than one million dollars for the reprints of a single article.[53] No wonder the drug industry sponsored studies have more impact than those sponsored by government.

Having an impact is highly sought after by journals. Impact is the factor academics use to assess the eminence of a journal. The impact factor is a measure of how much on average research papers in a given journal are cited or referenced by other research papers. Research published in 2009 found that ninety-two government-funded studies averaged 3.74; fifty-two studies funded entirely, or in part, by industry had a much higher average impact factor of 8.78.[54] What that signifies is that studies financed by drug companies are much more likely to appear in the more highly circulating and respected journals.

Being published in higher-impact journals has at least two very important advantages: the research is to be viewed as of superior quality; the article is more likely to be read. Having so much money to spend helps sell even more drugs. The pharmaceutical industry is easily able to purchase a half billion dollars worth of advertising space a year in academic journals, expenditures that often are the single largest source of income for these journals. Since academic journals are so dependent on this revenue from drug advertisements, they serve as an incentive to please drug companies, to try to avoid going against their interests, and even to recommend their products.[55]

Consider, for example, what happened after the Annals of Internal Medicine in 1992 published an article critical of drug advertisements that appear in academic journals. The article found that 44% of those drug ads are written in such a way that doctors who had not other sources of information would be led to prescribe improperly. Furthermore, the article indicated that 92% of these ads were, in certain respects, violating FDA rules.

The pharmaceutical industry responded by withdrawing many of its ads. To avoid such financially adverse events, editors of medical journals practice the self-censorship necessary to avoid arousing the ire of their major advertisers. This last observation was attributed to Dr. Marcia Angell, a former editor of the NEJM.[56] She surely knows what is at stake: the NEJM is alleged to receive about 10–20 million dollars annually from its drug company advertisers.[57]

Income from drug advertisements do influence what is published in academic journals. A study in 2011 of all the issues of eleven journals read by general practitioners found 412 articles making drug recommendations. Journals issued free of charge and subsidized by drug ads virtually always recommended the use of drugs in questions: journals whose revenue came through subscription fees did not for the most part recommend the use of these same drugs.[58]

---

[53] Abramson, Overdosed America. See also Goldacre, Bad Pharma, 306–307.
[54] Goldacre, Bad Pharma, 307–308.
[55] Ibid, 304–306.
[56] Abramson, Overdosed America, 113.
[57] Goldacre, Bad Pharma, 305.
[58] Ibid, 306.

For academic journals to prefer articles that feature studies carried out and written by drug manufacturers is at the same time to prefer studies that claim positive results over those obtaining negative results. These positive studies are also published more quickly, while negative studies are much more likely not to be published, or else only after considerable time elapses. Such a policy facilitates and protects a company's profits: it does so at the expense of people's health.[59]

An example of how devastating the delay in publishing negative outcomes can be involved a class I antiarrythmic drug (so-called) that were prescribed routinely into the 1990s. These drugs were effective for decreasing the frequency of extra heartbeats. However, they also increased deaths. The fact that these drugs resulted in higher death rates was discovered in research completed in 1980 but that was not published until 1993. An article appearing in JAMA in 2003 reported an estimated 20–75 thousand deaths occurred each year in the U.S. during the 1980s from inappropriately prescribing these class I antiarrythmic drugs. The JAMA article also noted that had doctors known about the deadliness of these drugs much earlier its use could have been stopped much earlier, saving many lives.[60]

The proclivity of medical journals to publish positive results contributes to the great harms often created by missing data. We reported earlier that Vioxx had caused 88– 138,000 excessive heart attacks before the missing negative data was disclosed.[61] In the same year that the sale of Vioxx was halted (2004), the International Committee of Medical Journal Editors, comprised of the editors of the world's most influential journals, pledged in a published statement to publish only those clinical trials that had been properly registered in advance of undertaking them and thereby put an end to publication bias.[62] As noted above, that did not happen; the editors simply did not comply with their own promised policy. Rather, they chose not to suffer the enormous financial losses of industry funded studies that are not preregistered.

Under the influence of drug company money, medical journals contributed to misleading doctors who rely on published data to guide them in choosing particular drugs to prescribe: "any attempt to recommend a specific drug is likely to be based on biased evidence."[63] That is the conclusion of a Swedish study of applications for five new antidepressant drugs submitted to the Swedish Drug Authority. The publication bias that results from missing negative data was very much in evidence.

When what is published in medical journals on which doctors rely for practicing scientifically and ethically sound medicine is biased, everyone loses: doctors, patients, taxpayers, and all the governmental and other organizations that participate in seeking good medical care. Physicians trust the leading medical journals and what they publish as Abramson discovered when he told a fellow physician about

---

[59] Abramson, Overdosed America, 113.
[60] Ibid, 114.
[61] Faloon, Pharma, 216–217.
[62] Goldacre, Bad Pharma, 51–52.
[63] Abramson, Overdosed American, 115.

the harmfulness of the drug Vioxx. That physician told him "I don't believe it," and showed no interest in looking up the sources Abramson offered for his assertions.[64]

What we have witnessed in this segment of our section on the influence of drug companies is just how extensively and successfully they wield their influence upon what physicians include in the array of medical knowledge they rely upon in their clinical decision-making; that influence is present wherever physicians can be found. The considerable money drug companies spend to make this happen is well worth the resulting increased sales and use of their products.

It is important to be fully aware of what it is that makes physicians vulnerable to this influence of drug companies. To ascribe that to simply the acceptance of gifts, amenities, and money does not account for the fact that the overwhelming majority of physicians do not regard themselves as being influenced by drug company largesse. These physicians will understandably deny that they are being corrupted morally in any way. However, some of them are engaged in what they should know to be unethical conduct. How does that come about?

I will limit myself to two very significant examples. As noted, there are medical experts with financial ties to drug companies who are among those who set up guidelines for physicians specifying what counts as a condition or illness that should be treated with drugs. When these are drugs manufactured by companies with financial ties to experts involved in setting such guidelines, those experts should recuse themselves from the decisions and the process of decision-making in such cases. That tends not to be happening. Indeed, the FDA knowingly recruits medical experts to serve on their committees to approve new drug products with such conflicts of interest who consent to serve in that capacity. These failures to avoid biases is a failure to be aware of, or to follow, an essential criterion, discussed in Chap. 2, a criterion for making rationally justifiable moral judgments, namely impartiality. It is a hallmark of professional ethics. In a court of law, assuring impartiality requires judges with a conflict of interest in a given case to recuse themselves; physicians have the same moral obligation in situations like those illustrated above.

Failure to meet these moral obligations have serious adverse consequences: in the case of setting guidelines, overdiagnosing and overmedicalizing, and thus exposing evermore millions of individuals to unnecessary harms without benefiting them; in the case of the FDA approval process, increasing the acceptance of far too much rigged, unethical and scientifically flawed research, and thus exposing many individuals to non-beneficial drugs that can only harm them.

How conflicts of interest should be handled in a morally justifiable way, and what situations should be avoided entirely because failure to do so will mean one cannot avoid a conflict of interest, is examined in detail in the next chapter.

---

[64] Ibid, 203.

6.1.4 Deceptive, Illegal, and Harmful Drug Advertisements

On the face of it, there is nothing remarkable about the fact that drug companies, like other commercial enterprises, promote their products and generally do so by means of advertisements. Why then would the European Union ban pharmaceutical companies form direct-to-consumer advertising, and why are the U.S, New Zealand, Pakistan, and South Korea the only countries permitting it?[65] There are a number of significant reasons. For the sake of achieving justice in the provision of medical care, advertising, as most of it is currently practiced by drug companies cannot be tolerated. The reasons for that we will now document.

6.1.4.1 Disease Mongering

In Chaps. 4 and 5, we pointed out the pervasive and excessive overdiagnosing, over-treatment, and overmedicaliation that occurs when normal conditions are regarded as illnesses or precursors to illnesses and so in need of treatments with drugs: disease mongering by means of advertisements help foster these very costly and very harmful unwarranted practices. As the psychiatrist Frances has indicated: "only a very few people have severe mental illness, many more have mild mental illness, but the real mother lode of market share is the worried well."[66]

Let us trace how the drug Paxil, manufactured by Glaxo-Smith-Kliner, was advertised to illustrate a clear case of disease mongering. Paxil was approved in 1996 by the FDA for treating depression, a market quite saturated with a number of antidepressants. To create a market, makers of Paxil chose to specialize in treating anxiety. They sought and received approval from the FDA to treat social anxiety disorder (SAD) and generalized anxiety disorder (GAD).[67] Through their advertising campaigns they contributed to the medicalization of emotions such as worry and shyness. For Conrad, Paxil "is a key example of how pharmaceutical marketing can reframe and medicalize common human characteristics and experiences," and a case that "demonstrates how pharmaceutical companies are now marketing diseases, not just drugs."[68]

It is true that psychiatrists in DSM-IV identified SAD and GAD as mental disorders. However, these were narrowly defined with criteria that require anyone so categorized to have multiple, persistent symptoms that are associated with severe distress and impaired functioning.[69] Compare that description with one of the ads for Paxil that says: "imagine being allergic to people," and another proclaiming

---
[65] Goldacre, Bad Pharmacy, 247.
[66] Frances, Saving Normal, 93.
[67] Peter Conrad. The Medicalization of Society: On the Transformation of Hyman Conditions into Treatable Disorders (Baltimore, MD: The Johns Hopkins University Press, 2007), 17.
[68] Ibid, 19.
[69] Ibid, 17–18.

"Paxil's efficacy in helping SAD sufferers to brace dinner parties and public speaking."[70]

Recall what Paxil's product director had to say about what his company did with SAD:

> Every marketer's dream is to find an identified market and develop it. That's what we were able to do with social anxiety disorder.[71]

As Frances has bluntly observed, "the best way to sell psychiatric pills to sell psychiatric illnesses."[72]

This resort to disease mongering by drug companies has expanded to include young children, adolescents, and the elderly. Doing this glosses over the special difficulties of diagnosing those populations, and their extra vulnerability to harmful side effects: the excessive use of antipsychotics in nursing homes, for example, has led to excessive deaths.[73] Recall that adolescents are so prone to suicides and violent impulses that prescribing these drugs is not recommended in Great Britain.

Legal psychiatric drugs are also addictive, embracing 7% of Americans.[74] And increasingly responsible for accidental iatrogenic deaths.[75] Additionally, there is the stark and highly disturbing fact that approximately three quarters of the 11% of the U.S. population taking antidepressant drugs do not have symptoms of depression according to studies published in 2011 and 2012. This Frances attributes to the placebo effect since antidepressants are no more effective than placebos. Yet many continue using antidepressants to the tune of a significant portion of the 12 billion dollars spent annually on antidepressants. Frances concludes that all this money and all these risks of exceedingly harmful side effects are for many patients, "no more than highly advertised, oversold and very expensive placebos prescribed for a fake diagnosis."[76]

Additionally, so-called "direct to consumer ads" have a very specific deleterious effect upon the relations between patients and their doctors. The ads prompt patients to flood doctors with requests for specific drugs for symptoms and ailments the ads have depicted. Frances speaks of "an army of patients primed by advertising to ask the doctor for a pill," and the doctor as one who has received much of his education from drug company salespersons who also provided the doctor with free samples.[77] Abramson refers to the success with which advertisers can "evoke emotional responses that are strong enough to override traditional doctor-patient relations."[78]

---

[70] Ibid, 18.
[71] Ibid.
[72] Frances, Saving Normal, 92.
[73] Ibid, 95.
[74] Ibid.
[75] Ibid, 106–107.
[76] Ibid, 100–101.
[77] Ibid, 102.
[78] Abramson, Overdosed America, 155.

Does this effort work? To that question, Abramson cites three studies: one revealed that doctors prescribe drugs requested by patients 50% of the time; another study found that doctors do so 75% of the time; yet another study discovered that doctors prescribe what patients ask for 80% of the time.[79] Abramson tries to persuade his patients that what the ads promise is not to be trusted and it not beneficial for health. So powerful is the influence of these ads that he does not always succeed. What Abramson laments are the instances when he is not able to move the discussion beyond whether or not the latest drug should be used and steer the patient toward more efficacious means to alleviate symptoms and maintain health by avoiding certain substances, like allergens, and by developing a more active, fitness promoting, way of life.[80]

6.1.4.2 Downplaying Side Effects

When the FDA first allowed direct-to-consumer advertising by drug companies, these were tightly regulated. After much pressure from drug companies, the FDA loosened the rules so that TV and radio ads could include an account of the condition the drug was designed to treat and do so without indicating all the side effects. The ads can direct the viewers or listeners to a specific magazine for more information. For example, Abramson cites a T.V. ad for Zoloft, an antidepressant, that first dissipates the clouds of depression and then briefly flashes on the screen: "see our ad in Slope magazine." Abramson doubts very many viewers will spend time looking for Slope at newsstands, particularly those who are depressed.[81] Drug ads promote the idea that the right drugs are necessary for one's health and happiness.[82] Keep in mind that you would be less likely to buy into the idea if the ads highlighted the side effects and contra indications—all of them!

One of the most egregious instances of omitting references to the side effects of drugs was certainly a major factor in allowing Vioxx to reap big profits before Merck took it off the market. There is the following e-mail from the head of Merck's research department to his leading scientists when talks with the FDA were not going as he wished:

> Twice in my life I have had to say to the FDA, "that label is unacceptable, we will not under any circumstances accept it."… I assure you I will not sign off on any label that has a cardiac warning for Vioxx.[83]

Though the FDA knew about the serious, extensive cardiovascular risks such as heart attacks and strokes, the agency yielded to pressure and permitted these side

---

[79] Ibid, 156.
[80] Ibid, 156–157.
[81] Ibid, 150–151.
[82] Ibid, 151.
[83] Faloon, Pharma, 220–221.

effects to be left off the warning box. Months later this information was included in the label's "precaution box," normally so lengthy that few would read it.[84]

6.1.4.3 Illegal Practices

The unrelenting quest for expanding markets is expressed in advertising drugs for uses that have not been explicitly approved. Drugs are approved by the FDA on the basis of studies designed to prove a given drug is sufficiently effective and safe for a specific use. Whereas doctors may resort to prescribing a drug for uses that are "off-label," it is illegal for drug companies to try to get them to do this.

Frances presents us with a "Drug Company Hall of Shame" that lists thirteen notable instances of fines and settlements for off-label promotions and, in one case, fraudulent misbranding that caused false claims to be filed against the drug OxyConton. The biggest fine listed was three billion dollars; the one for fraud was 635 million dollars; and most other fines were less than a billion dollars.[85] Frances concludes his reflections on these practices by noting that: "only much bigger fines and tighter regulations can tame this beast."[86]

Does the reader need more reasons to understand why it is that the EU bans direct-to-consumer advertisements, and why so very few countries—four—permit them? And I assure you what I have presented is only illustrative, not exhaustive.[87]

6.2 Unjustifiable and Unjustly High Prices for Drugs, Medical Devices and Medical Care

The pharmaceutical industry is very successfully lucrative, with sales of over 700 billion dollars a year worldwide: half of these occur in North America and one fourth in Europe. That yields drug companies a profit margin of 17%.[88] Such incredibly high profitability is three times the average achieved by other Fortune 500 companies.[89]

Drug companies defend their high prices by pointing to their support for research and improving patient care. However, they spend 60 billion dollars for promotion, twice what they spend on research.[90] Furthermore, these drugs do not always

---

[84] Ibid, 221.
[85] Frances, Saving Normal, 96.
[86] Ibid, 95 and 97.
[87] For example, see the account given by Goldacre, Bad Pharma on advertising, 247–271 and the even longer account of marketing, 240–340, that includes material we have covered earlier in this chapter.
[88] Frances, Saving Normal, 89–90.
[89] Abramson, Overdosing America, xiii.
[90] Frances, Saving Normal, 90.

improve patient care. The research they sponsor is often carried out, analyzed, and published in ways that favor and promote drugs that are neither efficacious nor safe enough to [qualify as improvements] to patient care.

We turn now to consider some of the reasons high prices occur and why they cannot be justly defended.

6.2.1 Governmental Facilitation and Protection of High Drug Prices

To begin with, drugs licensed by the FDA are patented. A patent grants the maker of a licensed drug exclusive rights to make and sell that drug for 20 years from the time the drug was filed. As we have noted previously, the prices of patented drugs in the U.S. are priced 50% higher on average than the same drugs are in the major industrial countries.[91]

This disparity in prices is being protected by the U.S. government. First, by its refusal to negotiate prices it pays for the drugs covered by Medicare and Medicaid in the Affordable Care Act. Secondly, prices are protected by the governmental agency, the FDA, by banning cheaper imports from other countries, except by individuals for their own use. Without evidence, the FDA declares them to be unsafe even though they are identical to the drugs being patented in the U.S. In fact, most of the drugs sold in the U.S. have in them active ingredients that have been synthesized in the very countries that the FDA labels as untrustworthy.[92]

The current high prices are unnecessary. The truth is that drugs would still be profitable at lower prices than those that obtain in Europe. The difference between the cost of active ingredients and the price charged for the pills that contain them is huge. For example, the widely prescribed drug Lipitor sold for $272.37 per 100 tabs/capsules in 2002: the active ingredient cost $5.80; that constitutes a mark up of 4686%. In 2014, the price of the same amount of pills more than tripled to $823.09. The markup for the drug Xanax was a whopping 569,958% in 2002, and still the price for 100 pills in 2014 rose just over two and a half times what it was in 2002.[93]

Someone might ask about the cost of labor. That cost is kept to a minimum by using laborers in China and India in drug-making factories approved by the FDA. Though the FDA bans cheaper imports, it does not do so for drug companies. Furthermore, in the case of China, the FDA is in no position to declare these imports

---

[91] For a comparison of U.S., European, and Canadian drug prices see Faloon, Pharma, 280. In this sample of widely prescribed drugs, some U.S. drugs sell for more than six times what they cost in Europe; none sell below double what they cost in Europe.

[92] For a detailed discussion of the FDA's policy on the importation of drugs, see Faloon's Pharma, 275–287.

[93] There is a list with similar information for 14 additional drugs provided by William Faloon in his article "how to Turn 8 Pennies into 600 dollars," in the September, 2014 issue of Life Extension, on page 11.

to be safe since the FDA can only inspect these drug-making factories every 13 years. Indeed, defective ingredients from India and China are sometimes to be found in prescription drugs sold in U.S. pharmacies.[94]

The FDA not only protects high drug prices but also facilitates achieving them. Albuterol is the name of a commonly prescribed asthma medication. The price of it rose from 15 dollars to 100 dollars, a more than sixfold increase, when this off-patent medication went back on patent.[95] How this came about is another egregious example of unethical and unjustifiable conduct made perfectly legal by the FDA's submission to lobbying by the companies that have lost patents.

This is how the deed was done. In order for the asthma drug Albuterol to be inhaled effectively and efficaciously a propellant is essential. All Albuterol drugs used CFC (chlorofluorocarbon) as the propellant. Pharmaceuticals that lost patents for medications that depended on CFC did not want to lose their monopoly on such a profitable market. To the tune of 520 thousand dollars, they set up a group to lobby the FDA to disallow the use of CFC in all drugs in order to protect the environment. CFC is an ozone-depleting agent that had been banned from use in a number of household appliances and industrial equipment but permitted in small amounts. Drugs like those in asthma inhalers were therefore exempted until late 2013. In the meantime, these same pharmaceutical companies developed patented combinations of Albuterol and other inhalants that contained propellants other than CFC.

The lobbying to ban CFC proved successful. In 2005, the FDA approved a ban on many CFC based inhalers to begin in 2009: by late 2013 the ban extended to all drugs that contained CFC. This was done despite the views of some scientists that the tiny amount of CFC in inhalers did not have a significant negative impact on the ozone layer. The result of the FDA capitulation to the drug lobby was to put an inferior medication on the market with a monopoly that left them free to raise prices. And raise them they did from 15 dollars to 100 dollars. After carefully documenting this utterly unethical sought and unethically obtained monopoly for pricing the asthma drug Albuterol, Faloon asserts that:

> Schemes like this to rip off consumers are not exceptions. They are customary practices of companies that routinely deceive courts, Congress, and the FDA to deny genuine competitors access to the market and stomp out the introduction of new medical products that could save lives and lower healthcare costs.[96]

One can apply what Faloon is saying to the whole range of practices we have been discussing so far in this chapter, namely the deceptive research, publications, advertising, and "medical education" carried on by drug companies. Hadler characterizes the overall purpose of all these activities as the pursuit of the "blockbuster:"

---

[94] Faloon, Pharma, 147.
[95] William Faloon, "Collapsing Within Itself," Life Extension, August, 2014, 9. The title refers to the financial collapse of the whole U.S. healthcare system.
[96] Ibid. See also Melody Peterson, Los Angeles Times, "Free coupons help drug makers hike prices by 1,000%," in New Hampshire Sunday News, December 25, 2016, B1–B2.

Since for the sake of science has no place in this pursuit... To be a blockbuster demands convincing a large population that adverse events are a price worth paying. To be a blockbuster demands a license from the FDA that does not squelch the exercise.[97]

And, according to the evidence we have presented, many doctors and patients are among those who are persuaded that these drugs are worth the cost and the risks of their harmful side effects. As a result of the kinds of drug company practices we have covered, the cost and use of drugs is exorbitant, unjustifiably and unsustainably so.

6.2.2 Hospitals: Charging Highly Inflated Prices for Drugs, Medical Devices and Every Other Aspect of Medical Care

Faloon is one of a number of physicians who have long been critical of the unnecessarily high prices that drive up the cost of medical care. However, the prices experienced by a woman named Jeanne Pinder were of such magnitude that he professes that they "even startled me."[98] Ms. Pinder understandably questioned being charged 1419 dollars for an anti-nausea drug (ondansetron). She found out that ondansetron could be had for 2.49 dollars at a local drug supplier: the hospital's price was 569 times as much! Given a price like that, Ms. Pinder wondered what insurance would pay for that same drug. For what was billed at 1419 dollars by the hospital, what she discovered was that the Veterans Administration would pay 15.76 dollars, Michigan Blue Cross Blue Shield 17 dollars, and Medicare 24.36 dollars. Consider these payments are for a drug that can be purchased for less than 3 dollars.

In 2013, the New York Times carried an article that investigated what hospitals were charging patients or their insurers for saline IV solution.[99] What turned up on some of the patients' bills were prices 100–200 times more than what the manufacturers charged. Patients were charged additionally for the administration of the IV solution.

Manufacturers are required by the federal government to report their prices to them. Medicare pays the national average price for a product plus 6%. So in 2013, Medicare would pay 1.07 dollars for a one liter bag of normal saline. Even so a New York hospital charged a private insurance company 91 dollars for a bag of saline that the hospital purchased for 86 cents!

Hospitals all too often charge so much for their services that people's insurance does not nearly suffice to defray their costs. Many individuals and families are left to pay astronomical amounts for hospital services. An article in Time magazine,

---

[97] Hadler, The Citizen Patient, 62.
[98] Faloon, "Collapsing Within Itself," 8.
[99] Ibid.

published in 2013, documented a number of such instances.[100] I have selected two representative examples for illustrative purposes.

This is what happened to the couple bearing the pseudonyms Rebecca and Scott. When Scott began violently gasping for breath, Rebecca anxiously drove him to the emergency room at the University of Texas Southwestern Medical Center.[101] Both Rebecca and Scott feared for his life. They were not thinking about any costs they would incur. What they could not know or even imagine is that the bill they would receive for treating Scott's pneumonia in the hospital for 32 days would far exceed what their insurance would pay: it was 161 pages long and 474,064 dollars high! Their insurance brought down the total charge for them to 402,955 dollars.

Here are some of the costliest items on the hospital's bill: Scott's room, just short of 2300 dollars a day totaled 73,376 dollars, for "resp services" 94,709 dollars, that is, for supplying Scott with oxygen, checking his breathing, and supervising [oxygen inhalation] made up of multiple charges of 134 dollars a day for which Medicare would pay 17.94 dollars; and for "special drugs," most of which were no more special than "sodium chloride .9 percent," 108,663 dollars. This standard saline solution, injected intravenously to maintain Scott's water and salt levels, sells for 5.16 dollars per liter, bagged and ready for use; the hospital charged 84–13 dollars per bag for dozens of them.

Incomprehensibly, each of the charges above are topped by the 132, 303 dollars billed for "laboratory." That included hundreds of blood and urine tests, each pegged at 30–333 dollars. Medicare sometimes pays nothing for such tests, treating them as covered by the room free or, when they pay, 7–30 dollars per test. No executive at Texas Southwestern Medical Center, not even the president, paid $1,244,000 a year, made themselves available to explain their billing practices.

What is also left unexplained is why so many tests were done. A physician who consulted the Obama administration on healthcare policy regards about 60% of the labs being done to be unnecessary. Many are done continually to provide data for pedagogy associated with going on daily rounds.[102] Hospitals with in-house labs generate about 60% of all testing revenue and serve, therefore, as vital profit centers.

Although testing is a big source of profits for hospitals, profits are sought at every level. For example, the hospital indulged in triple billing for services in the I.C.U. First they charged more than 2000 dollars a day for being there with all its special equipment and personnel. To that they added 1000 dollars for a kit that gives the occupant a transfusion or oxygen. Yet there's still more to pay for each tool or bandage or whatever else is utilized. That is triple billing.

This practice was identified by Patricia Palmer a billing advocate hired by Scott and Rebecca. Palmer worked hard and managed to persuade the hospital to change the bill from 402,955 dollars to 313,000 dollars. At the time this was published, the

---

[100] Steven Brill, "Bitter Pill: How Outrageous Pricing and Egregious Profits are Destroying Our Healthcare," Time, March 4, 201, 16–55.
[101] Ibid, 38–40.
[102] Ibid, 40.

best offer the hospital would make was to cut the bill to 200,000 dollars if it was paid immediately, or to allow the 313,000 dollars to be paid in 24 monthly payments. This hospital bill does not include thousands of dollars in doctor bills, and 70,000 dollars owed a second hospital when Scott had a relapse. Rebecca sums up their situation this way: "We thought we were set but now we're pretty much on the edge."[103] What was described as their comfortable semiretirement has become uncomfortably tenuous. An insurance policy with a 100,000 annual limit is no match for the prices charged by many highly profitable nonprofit hospitals.

When 42-year-old Sean Recchi of Ohio was diagnosed with non-Hodgkins Lymphoma, his wife Stephanie elected to take him to MD Anderson Cancer Center at Texas University in Houston where her father had been treated successfully. They had an insurance policy that paid 2000 dollars a day for hospital expenses. The hospital would not accept what they considered to be their inadequate insurance coverage. Instead, Stephanie was told to send 48,000 dollars in advance to cover a treatment plan and 35,000 in advance for urgently needed treatment, a total of 83,000 dollars paid by her mother.

Overpricing was evident at MD Anderson in even the smallest matters. Imagine charging 1.50 dollars for one tablet of Tylenol when 100 identical tablets can be purchased on Amazon for 1.49 dollars. A chest x-ray, a service reimbursed by Medicare for 20.44 dollars, was billed for 283 dollars. Then there's one dose ([660 mg]) of Rituxin at a cost of no less than 13,702 dollars! On average, this dose is sold to hospitals for 4000. MD Anderson professed to be in line with other major hospitals and academic centers in charging this much.

MD Anderson is a nonprofit hospital. How necessary then are the prices it is exacting? Not very! Their profits amounted to 531 million dollars in 2010, 26% of 2.05 billion dollars in revenue. And they are this profitable while paying their president, 1,845,000 dollars that year.

If MD Anderson adhered to the Hippocratic ethic, they would have a great deal of money they could afford to share with Sam Recchi who is underinsured for the kinds of exorbitant prices they charge. Instead they even charged 13,702 dollars for a drug that could have been offered free of charge. How so? MD Anderson did not inform Sean of this possibility. He and Stephanie only learned of this possibility after they left MD Anderson because they could not afford their prices. Sean's doctor in Ohio informed him that he could receive Rituxin free and that is what happened.[104] Happily, Sean was successfully treated.

Rituxin is a product of Biogen Idec that it makes and sells in partnership with Genentec. Rituxin is made, tested, packaged and shipped to MD Anderson for 300 dollars a dose and sold to them for 3000 dollars to obtain a profit for a dose of Rituxin. That Genentec has a charity program is commendable. However, what they have donated since 1985 amounts to less than 1% of Genentec's sales.[105]

---

[103] Ibid.
[104] Ibid, 36.
[105] Ibid.

Cancer drugs are exceedingly profitable drugs. We remind the reader that much money (tens of billions annually in the U.S.) would be saved, and drug companies would still be profitable, if prices for drugs were negotiated as they are in other countries.

You may think that major hospitals have to charge what they do! Think again. Among the top ten largest nonprofit hospitals, University of Pittsburgh Medical Center ranks first in profitability with an operating profit of 769,704,054 dollars and first in what they pay their CEO, 5,975,462 dollars.[106] Norton Hospital, Louisville, KY is the least profitable of the ten nonprofit hospitals in question with an operating profit of 118,101,911 dollars and a CEO making a salary of 2,206,401 dollars. This ranks ninth for CEOs in this group of hospitals: that is almost four times the $561,210 dollars received by the CEO of the nonprofit American Red Cross.

With the prices being charged by some large hospitals, particularly cancer centers, it is no small wonder that 62% of personal bankruptcies involve illness or medical bills, and that 69% of those who went bankrupt for medically-related reasons had health insurance at the time they filed.[107] The Affordable Care Act laudably extended insurance coverage to a greater number of individuals and families. Nevertheless by not directly addressing the high cost of drugs, medical devices, and hospital services, far too many individuals and families with health insurance will not be able to cope with their medical expenses. At the same time, insurance companies are being forced to raise their premiums and deductibles, making insurance increasingly too expensive for businesses, individuals, and families, and for federal and state governments.

Let's briefly illustrate other kinds of unnecessary unsustainable pricing. Note just a few procedures. In the U.S. the average charged for a CT scan is 510 dollars; in Switzerland 319 dollars; in Canada, 122 dollars; and in India, 43 dollars. Appendectomies in the U.S. are done for an average 13,003 dollars, in Switzerland, 5640 dollars; in Canada, 5606 dollars; and in India, 254 dollars. A coronary bypass can be had in the U.S. for an average 67,583 dollars; in Canada, 40,954 dollars; in Switzerland, 25,486 dollars; and in India 4,525 dollars.[108]

Medical devices are unbelievably high priced and profitable for their manufacturers. However, even more incomprehensible are the prices exacted by some hospitals in the U.S. For example, Mercy Hospital in Oklahoma City, a nonprofit, tax-exempt charity and part of a very profitable chain, billed a patient 49,237 dollars for a Medtronic stimulator, implanted to ease back pain. That netted the hospital a profit of more than 150%, having paid 19,000 dollars for the stimulator.[109]

---

[106] See Ibid, 31 for these data for all of the top ten largest nonprofit hospitals. These data were the most recent available in 2013.

[107] Ibid.

[108] See Ibid, 30 for these price comparisons and comparisons with six additional countries charging considerably less than U.S. hospitals as well. On the same page some drug prices in the U.S. are compared with much lower prices in other countries.

[109] Ibid, 32–34.

It should be noted that Medtronic, the company that manufactured the stimulator, is very profitable.[110] In its October, 2012 quarterly report, Medtronic indicated that all of its spine products and therapies, that would include the Medtronic stimulator, cost 24.9% of what they sell them for. That's a 75.1% gross profit margin. Knowing how highly profitable device makers are, the Obama administration officials imposed a 2.5% tax on the sales of these devices as well as other technology such as CT scanners. The reason for this was to cover some of the cost for the subsidies provided by the provisions in the Affordable Care Act.[111] But the expansion of insurance coverage will no doubt mean that more devices will be sold and insurance companies will experience ever more pressure to increase their premiums and deductibles because the Affordable Care Act does not address the prices of these devices and technologies by imposing this tax. And as we have indicated earlier, there will still be many patients with insurance who can ill afford the hospital prices for the uses of medical devices and technologies, and some who will not be able to afford them at all.

Our segment on what some hospitals charge for their services is only illustrative, not complete by any means. Our aim was to document still more examples of the enormous amount of money in the U.S. medical system that can be saved and ample profits being gained that could and should be shared with patients: I use the word "should" because doing so is a matter of justice embodied in the traditional professional ethic; no one should be denied needed medical care who cannot entirely or at all afford it. Alas the current prices for drugs, medical devices, and medical services create so many extremely difficult obstacles for the many morally conscientious caregivers who strive to practice in accord with the traditional, Hippocratically inspired, medical ethic.

6.3 The FDA: Documenting Some of Its Inadquate, Biased, Unethical, and Illegal Practices

The Food and Drug Administration (FDA) was created for very good reasons. Included in its written mission statement are the following important goals and responsibilities:

> The FDA is responsible for advancing the public health by helping to speed innovations that make medicines and foods more effective, safer, and more affordable; and helping the public get the accurate, science-based information they need to use medicine and foods to improve their health.[112]

These are surely highly valued goals, the attainment of which, we can applaud. The safety, effectiveness, and affordability of our medications and food are obviously

---

[110] Ibid, 34–36.
[111] Ibid, 34.
[112] Faloon, Pharmacracy, 216.

essential for our wellbeing, health, and our very lives. If justice and sustainability in the provision of medical care is to be attained, the role of the FDA is pivotal. Tragically, however, what we have learned so far in this chapter and the previous two about the policies and practices of the FDA, is that many of them are not in accord with its mission; they violate rather than actualize justice. We begin by listing some of these as follows:

Knowingly approving and failing to withdraw seriously harmful drugs, the side effects of which can and do cause debilitating conditions and even death;

Knowingly withholding important information contained in negative studies that would reveal evidence of the extreme harmfulness and absence of beneficial effects of drugs receiving approval for being marketed and patented;

Knowingly permitting the omission of some of a drug's harmful side effects from advertisements and labels or otherwise making it difficult to find or notice certain side effects;

Knowingly approving drugs of little or no benefit that have not been compared to existing beneficial drugs nor with more effective therapies that do not require drugs at all;

Knowingly facilitating high prices by approving and thus paving the way to patients for drugs of little or no benefit when more effective, less costly, and safer drugs or non-drug therapies without side effects exist to treat the conditions these patented drugs are designed to treat;

Knowingly protecting high prices for drugs in the U.S. by blocking importation of identical, cheaper drugs; knowingly allowing the persistence of flawed and deceptive research and analyses of research on the part of drug and medical device companies that the agency could expose and stop.

There is at least one more policy of the FDA that is distinctly at odds with its mission and it is being carried out illegally: the FDA suppresses scientific evidence cited on behalf of the health benefits of food and nutritional supplements.

In 1999, a Federal Appellate Court declared the rules the FDA used to suppress the following four health claims to be unconstitutional: consumption of antioxidant vitamins may reduce the risk of certain kinds of cancers; consumption of fiber may reduce the risk of colorectal cancer; consumption of omega-3 fatty acids may reduce the risk of coronary heart disease; 800 mg of folic acid in a dietary supplement is more effective in reducing the risk of neural tube defects than a lower amount in foods in common form.[113] The Court found the FDA's health claim standard to be "arbitrary and capricious," so subjective that it does not allow anyone to specify the level of scientific evidence the FDA required for approving a claim. The FDA was ordered to change its standards so that it would have to prove with empirical evidence that a health claim was misleading and would not be rendered non-misleading by adding a disclaimer. The FDA was ordered to implement the Court's decision fully, faithfully, and right away. The FDA did not comply.

---

[113] Ibid, 315–318.

FDA's non-compliance with the Court's order was evident when, once again, it suppressed scientific evidence and was taken to court.[114] Blatantly, it disobeyed the 1999 Federal Court's decision by seeking to ban one of the very claims it had been ordered to allow, namely that 800 mg. of folic acid more effectively reduces the risk of neural tube defects than a lower amount found in foods. The FDA's ban was appealed to the U.S. District Court of Columbia. In 2001, this Court found that the FDA had acted unconstitutionally and did so in violation of the Federal Court of Appeals. Furthermore, the FDA was cited for ignoring solid scientific evidence for the efficacy of 800 mcg of folic acid supplements to prevent neural tube defects (NTDs) and thereby causing irreparable harm to about 2500 babies annually; most never attain adulthood, and those who do, are afflicted with serious handicaps. Also, this violation of the public interest is very costly; more than 500 thousand dollars over a lifetime for spina bifida, the most common NTD; Social Security payments of more than 82 million dollars annually.

Unbelievably, these court orders have not stopped the FDA from suppressing scientific claims made on behalf of foods. In 2005, the FDA moved to ban the sale of tart cherries. That it did despite the fact that numerous scientific studies have documented that tart cherries have some very significant health benefits.[115] These include reducing the risk of colon cancer and alleviating arthritis inflammation and pain, doing so more effectively than FDA approved drugs for that purpose.

Yet the FDA wrote to those who market tart cherries that they would be engaging in a criminal activity if they continued informing consumers of the scientific data lauding the health benefits of tart cherries. The FDA asserted, in their warning to twenty-nine companies that market cherries, that these claimed health benefits cause their product to be a drug that is not recognized as safe and effective when used as labeled. As a new drug, it cannot be legally marketed in the United States without obtaining FDA approval.

These astounding and utterly absurd assertions are made without denying the veracity of the scientific information offered by those who market tart cherries! And the FDA has not stopped this indefensible behavior. A more recent example involves walnuts. The claim made for walnuts is that eating them protects one against coronary heart disease. For this claim and the health benefits of walnuts there is considerable scientific evidence.[116] The FDA's letter to Diamond Foods that sells walnuts contains the same kinds of accusations and threats of criminal prosecution as those made against those marketing tart cherries.[117]

What are the effects of suppressing scientific studies that uncover what ailments various foods and supplements can effectively prevent, ameliorate, or cure? In all too many instances that means that less effective, more costly, and quite harmful drugs are left in place to be freely marketed with FDA approval, an advantage in

---

[114] Ibid, 315–322.

[115] Ibid, 214–216. The evidence for the health benefits of cherries; see footnotes in Ibid, 222–223.

[116] Ibid, 29–32. There is ample evidence for the health benefits of walnuts: see Ibid, 40–42, for footnotes 1–31.

[117] Ibid, 31–32.

marketing denied food and nutritional supplements. Surely it is unnecessary to point out that the FDA is exceedingly acting in defiance of its mission and even of constitutionally protected rights. I find such illegal and merciless disregard for the wellbeing of patients, and of all of us as taxpayers, by a government agency extremely hard to fathom.

What could possibly fuel these flagrant breaches of ethics and law? Conflict of interest. The physician Norton Hadler unequivocally asserts that the health care system in the U.S. is ethically compromised and that there is no debate about that.[118] What he sees being debated are: the degree of compromise; and more heatedly, what to do about it. He regards the debate as surrounded by clouds of "obfuscation" amid considerable finger pointing. In his view, what is hidden in these clouds is "an essence that renders the U.S. health care system essentially bankrupt. That essence is conflict of interest."[119]

In the FDA, the conflict of interest is very evident and the results of it very predictable. A study of a long series of FDA votes cast in the approval process found that those participants in this process with financial ties to a company were somewhat more likely to vote in that company's interest.[120]

As we have amply noted, the FDA all too often approves and fails to withdraw drugs that are known to be harmful and even fatal. One instance of this was sharply rebuked in an editorial appearing in the medical journal Lancet in May 19, 2001 and cited therein for conflict of interest.[121] In November 2000, the drug company had withdrawn their drug Lotronex after it led to the death of five patients and caused other serious side effects. Quite unjustifiably, the FDA was, at the time of the editorial seeking to reintroduce this drug. This prompted the author of the article to assert that:

> This story reveals not only dangerous failings in a single drug's approval and review process but also the extent to which the FDA, its Center for Drug Evaluation and Research (CDER) in particular, has become a servant of industry.[122]

The editorial calls attention to the hundreds of millions of dollars the FDA receives from industry. Indeed, the FDA supplements its federally supported budget by charging drug companies hefty fees for the processing of their "new Drug Applications" and in recruiting advisors in that process who have industry ties.[123]

The editorial points out that the FDA knew of the drug's risks from its own scientists before the drug was withdrawn by its manufacturer. Yet it decided to simply issue a warning, one that its scientists said was impractical and would not work for patients and physicians. The FDA officials dismissed the concerns of the scientists who proved to be right. The scientists who raised these issues not only felt intimi-

---

[118] Hadler, Citizen Patient, 1.
[119] Ibid.
[120] Goldacre, Bad Pharma, 126. See further studies of conflict of interest on page 127.
[121] Faloon, Pharmacy, 307–309.
[122] Ibid, 307.
[123] Hadler, Citizen Patient, 87.

dated by their senior colleagues but were also subsequently excluded from further FDA discussions about the future of Lotronex.

The Lancet had published some of the trial data on which the FDA based its approval of the drug. The editor wondered what happened once the adverse effects became known. That led to uncovering how the FDA deals with science and how fatally its independence is compromised by being financed by industry. In this case, the scientists were simply ignored.[124]

The conflict of interest is recognized and deplored by some working within the agency. The Associate Director for Science and Medicine in the Office of Drug Safety plainly informed the Committee on Finance in the U.S. Senate that "the FDA has become an agent of industry. I have been to many, many internal meetings, and as soon as a company says it is not going to do something, the FDA backs down. The way it talks about industry is 'our colleagues in industry.'"[125] What should be done about the conflict of interest that exists within the FDA, and about the rather extensive failure to carry out its mission, will occupy us in our next and last chapter.

6.4 Defensive Medicine, Malpractice Insurance, and Litigation

Reducing the costs of defensive medicine, the premiums of malpractice insurance, and litigating malpractice suits would help make the costs of medical care in the U.S. sustainable. A study of all of these costs combined found, that in 2008 dollars it was 55.6 billion dollars or 2.4% of the total spent on health care in that year.[126] Of this total, the study estimates that defensive medicine cost 45.59 billion dollars.[127]

It is difficult to determine the extent of defensive medicine. Still three-fifths of physicians in the U.S. acknowledge that they engage in more diagnostic testing than they deem to be necessary because of their fear of litigation.[128] The study cited above may have yielded a very conservative estimate of the cost of defensive medicine. The incentives to do so are very evident: there are no penalties for ordering an extra test; there are for failing to do so.[129] Being sued successfully, especially when the suit should have been rejected in court, strongly inclines a physician to err on the side of doing diagnostic tests beyond those they consider necessary and sufficient. Welch discusses just one such sad instance experienced by a colleague and cites

---

[124] Faloon, Pharmacy, 308–309.
[125] Goldacre, Bad Pharma, 127–128.
[126] Michelle H. Mella, Amitabh Chandra, Atul A. Gawanda, and David M. Studdert, "National Costs of the Liability System," Health Affairs, September 2010, 29:9, 1569–1577.
[127] Ibid, 1570.
[128] Abramson, Overdosed America, 161–164.
[129] Ibid, 84. Welch, Overdiagnosed, 161 makes this same point.

some research showing that the practices of other physicians are similarly affected by suits against them.[130]

There is research that strongly suggests that defensive medicine is quite prevalent.[131] A survey asked physicians five questions that tested how fearful they were about possibly being sued for malpractice. The responses to two of these questions are particularly telling: 78% agreed, some strongly, that: "relying on clinical judgment rather than technology to make a diagnosis is becoming risky because of the practice of malpractice suits;" slightly more than 60% agreed, some strongly, that: "I order some tests or consulting simply to avoid the appearance of malpractice."[132]

There is a study that purports to show that curbing defensive medicine will not save nearly as much money as some have contended.[133] However, I do not see that the study had any effective method of specifying what constitutes defensive medicine. Nevertheless, they consider the cost of defensive medicine at 1.34% of medical care costs should be eliminated by some kind of tort reform.[134]

There is a serious problem that results from the high cost of malpractice insurance. Abramson calls attention to a rebellion among doctors brought about by the rising cost of malpractice insurance: they are opposed to have to "pay the price for our litigious culture (and a few bad doctors) regardless of their own track record and commitment to quality care."[135] While some are protecting their assets and practicing without insurance, there are others who are relinquishing the practice of medicine entirely. The loss of physicians is not only sad but it also comes at a time when these losses undermine the expansion of insurance coverage brought about by the Affordable Care Act; the U.S. currently has a shortage of primary care physicians and it grows worse as more people receive insurance.[136]

Chapters 4, 5, and 6 have presented a great number of practices within the whole system of providing medical care in the U.S. that drive up the cost of providing it. We have focused on practices that are unnecessary, scientifically unjustifiable, and harmful, and in other ways in violation of the demands of justice that characterized the ideals much adhered to by Hippocratic physicians and Hippocratically oriented physicians. These injustices we have identified render medical care in the U.S. unsustainable. Hence, our last chapter explores suggestions as to what is needed to achieve justice and sustainability in the provision of medical care.

---

[130] Welch, Overdiagnosed, 161–164.

[131] Emily R. Carrier, James D. Reshovsky, Michelle M. Mello, Ralph C. Hayrell, and David Katz, "Physicians' Fear of Malpractice Lawsuits Are Not Assuaged by Tort Reforms," Health Affairs, September 2010, 29:9, 1585–1592.

[132] Ibid, 1570.

[133] J. William Thomas, Erika C. Ziller, and Deborah Thayer, "Low Costs of Defensive Medicine, Small Savings From Tort Reform," Health Affairs, September 2010, 29:9, 1578–1584.

[134] Ibid, 1583.

[135] Ibid, 82–84.

[136] Collwell JM, Cultice JM, Kruse RL, "Will Generalist Physician Supply Meet Demands of an Increasing and Aging Population?" Health Aff (Millwood), 2008 May-June 27(3):[w] 232–41. Epub 2008 Apr 29.

Chapter 7
Suggesting Policies and Practices for Increasing Justice and Assuring the Sustainability of the U.S. Healthcare System

Abstract The goal of this chapter is to suggest policies and practices that would render the U.S. health care system just and sustainable by: (1) Curbing current unethical and scientifically unsound medical practices that are unjustifiably harmful and of little or no benefit; (2) Financing healthcare by drawing upon the profits earned by all those individuals, organizations, and manufacturers that provide healthcare. To have this happen, the healthcare system must be purged of conflicts of interest, especially the FDA. Dealing justly with costs requires that excess profits that amount to gouging be mandated to be used for medical interventions.

"If health care could take the high moral ground, the nation would be rewarded with better health at a tolerable cost."[1] That is what the physician Norton Hadler argued when asked for his advice by one of the intellectuals who helped to formulate the Affordable Care Act Congress passed on March 10, 2010. The reason he gave for his argument was that the Affordable Care Act "perpetuated all that was morally bankrupt about the American health care system."[2] For Hadler, that meant that the Affordable Care Act's efforts to bring down costs would utterly fail because it was ignoring the magnitude and costs of useless, scientifically unjustifiable, and hence unethical medical interventions. Hadler knew his counsel was not being heeded: indeed it wasn't.

What I have presented in Chaps. 4, 5 and 6, backed by abundant scientific evidence, supports the rest of Hadler's contentions as stated in the paragraph above. First of all, that the Affordable Care Act has left intact many, but not all, of the egregious, morally unacceptable practices of the U.S. health care system. Secondly, since this Act left standing, not only a great many useless, scientifically unjustifiable, ethically indefensible medical interventions, but also exceedingly high priced medical products, practices, and services, particularly in large hospital complexes, the Affordable Care Act is unaffordable. And thirdly, the impact of removing use-

---

[1] Norton M. Hadler, *Citizen Patient: Reforming Health Care for the Sake of the Patient, Not the System* (Chapel Hill, NC: The University of North Caroline Press, 2005), 179.
[2] Ibid.

less, unnecessary and scientifically unjustified medical interventions would not only achieve large financial savings, but also the avoidance of much harm and better health for the public.

Physicians have begun to subject their clinical decisions to careful scientific scrutiny for the very purpose of setting guidelines to refrain from tests and treatments that are unnecessary, more harmful than beneficial, or replaceable by less costly alternatives, and in these ways save considerable money. I refer to the organized efforts of the Choosing Wisely Movement described in Chap. 4. However, there are other key decision-making bodies needed to expand the scope of the efforts to eradicate what is unjust about the U.S. health care system. I have in mind the federal government, a properly reformed FDA medical schools, hospitals, and medical journals. In this chapter, therefore, I have suggested a number of policies and practices that illustrate what these entities, and physicians as well, have available to render the whole U.S. health care system more just and, at the same time, sustainable. These policies and practices are only suggestions. Ultimately, what works best needs to be determined and implemented by these participants in the U.S. health care system, and that is as it should be. I would, however, offer them this advice: their deliberations will only be fruitful if guided by the insight we owe to Robert Frost, that only mercy can make the unjust just.

7.1 What the Federal Government Can and Should Do to Achieve Justice in the Provision of Health Care

The current government's policies, actions, and failures to act contribute substantially to making health care in the U.S. ever more costly, unjust, and financially unsustainable. However, the government's involvement in the health care system is such that it is an excellent position to put its policies on a path that increases justice and decreases injustices within the U.S. healthcare system. The suggestions that now follow identify some ways the federal government can do just that.

7.1.1 Reform the FDA

We begin with the Food and Drug Administration (FDA) because putting it on course to fulfilling its mission would prevent much harm and many fatalities, and also save hundreds of billions of dollars now annually being wasted on useless and harmful interventions. As an agency created by the federal government to assure the safety of medicine and food, as well as helping to make them more affordable, the Federal government has the responsibility to assure that the FDA work conscientiously and effectively to meet these responsibilities spelled out in its mission statement. That mission includes working to see to it that there is accurate scientific information available for the public's scrutiny and use. Doing that is essential if the FDA is to fulfill its mission to advance the health of the public.

In Chap. 6, we identified the kinds of FDA policies and practices that cannot be morally and scientifically defended: the Federal government should have them discontinued; the suggestions that now follow will make it abundantly clear that the government can find ways to accomplish that.

7.1.1.1 Drug Approval Practices that Should Be Disallowed

One would think that it would not be necessary to specify what current practices of the FDA cannot be scientifically and ethically justified; the FDA certainly has the capacity and should have the desire to identify and stop them. However, as indicated in Chaps. 4, 5 and 6, the FDA all too often does not do that. Therefore, these kinds of unjustified activities need to be identified and halted by Congress, the body that should hold its agency accountable. The taxes funding the FDA should not be misspent, and above all, should not be a source of inflicting avoidable harms on those footing the bill.

To being with, the FDA has the scientific knowledge at its disposal to end its widespread, frequent approvals of drugs that are of little or no benefit even when compared to the effects of placebos used in randomized trials. Furthermore, trials for drugs seeking FDA approval should be compared with the best available, approved drugs at the time these trials are to be conducted. Such comparisons should not be limited to therapies that prescribe drugs since the most effective and efficacious interventions may involve a proper diet and physical activities. The FDA should be mandated to follow the Declaration of Helsinki and its ethically justified admonition that experimental subjects should not receive placebos in randomized trails when approved, scientifically supported therapies exist to treat the condition a newly created drug is being tested to treat. The FDA has chosen to bypass this clear demand of justice, and by doing so needlessly depriving experimental subjects of therapeutic therapies they may require. At the same time, this allows research to be conducted and approved that does not reveal whether that drug is more efficacious and safe than the best available treatment in use.[3]

There are two approval practices that definitely disregard what justice requires of us all: the vital moral responsibility not to harm, injure, and cause the death of

---

[3] Ben Goldacre, Bad Pharma: How Drug Companies Mislead Doctors and Harm Patients (London, Great Britain: HarperCollins Publishers, 2012), 131. The Declaration of Helsinki added an amendment that required compelling and scientifically sound methodological reasons for using a placebo in a trial testing the efficacy or safety of an intervention and then only if the patient receiving the placebo will not be subject to any risk of serious or irreversible harm. A study found that between 2000 and 2010 that 70% of drugs approved by the FDA that needed to be compared to existing treatments did so. Of these, one-third provided no evidence that compared the drug tested with the currently best available treatment. See also, John Abramson, Overdosed American: The Broken Promise of American Medicine (New York: HarperCollins Publishers, 2004), 102–103.

innocent individuals. For centuries physicians have regarded it as their duty to adhere to their understanding of this tenet of the Hippocratic ethic: first of all do no harm.

In Chap. 6, we called attention to the following two FDA practices that unquestionably violate this traditional medical ethic and its own mission:

> Knowingly approving and failing to withdraw some seriously harmful drugs, the effects of which can cause very debilitating conditions and, in some instances, even death; knowingly withholding information contained in negative studies that would reveal the evidence for the serious harmfulness and absence of beneficial effects of the drugs being approved.

Both of these practices took place in the FDA's handling of the drug company Merck's submission and marketing of the painkiller Vioxx. Unthinkable as it is, the FDA acts in these ways in the Vioxx case and in other instances like it even though their own scientists supply FDA officials with analyses that should lead them to avoid such ethically, scientifically, and legally unjustifiable practices.

Given the practices we have described, those in authority at the FDA should be held accountable for their actions in cases like the Vioxx case. As noted in Chap. 6, Congress heard testimony from the Director of the scientific segment of the FDA that 88–139 thousand excess heart-attacks were experienced by individuals who were prescribed and used Vioxx. For much less extensive harms car manufacturers are rightly subject to legal penalties for knowingly compromising the safety of their automobiles. FDA officials should be no less legally held accountable.

7.1.1.2 Tightening FDA Rules Governing Deceptive Ads

As things now stand, the FDA's prevention of deceptive ads is quite limited. The FDA does fine drug companies that tout using a drug for a purpose or condition for which the drug has not specifically been approved by the FDA.[4] However, even fines as large as the 1.3 billion dollars levied against Merck are easily absorbed by the highly profitable large drug companies and have not halted these deceptive ads. Therefore, some advocate much higher fines.

However, the FDA rules as presently constituted, permit ads to be deceptive and even more predictably harmful in other ways. Drug ads used to be restricted to marketing to doctors only. In 1985, the FDA allowed direct-to-consumer ads but tightly limited the content so that ads primarily promoted brand awareness and trips to the doctor's office. Responding to pressure from drug manufacturers, the FDA loosened its rules in 1997 by permitting TV and radio ads that describe the condition or conditions the drug was intended to treat, and to limit what side effects and contraindications have to be present in the ads themselves: audiences can be told to see a website or magazine for more extensive information. From 1994 to 2000 drug ads increased 40-fold.

---

[4] A list of some fines levied against drug companies appears in Allen Frances, Saving Normal: An Insider's Revolt Against out-of-control psychiatric diagnosis, DSM-5, big pharma and the medicalization of ordinary life (New York, NY: HarperCollins Publisher, 2013), 96.

When the drug companies pressed the FDA to sanction direct-to-consumer ads, they argued that the public should not be deprived of the information such ads would offer them. What is the public actually being told in these many ads? What studies have discovered is that drug ads usually do not provide what the public needs to know. Consider what researchers from Dartmouth Medical College discovered:

> Only 13 percent of drug ads in magazines used data to describe drug benefits; the remaining 87 percent relied on vague statements. Not a single ad in the study mentioned the cost of the drug. Only 27 percent of ads presented the cause of or risk factors for the disease, and only 9 percent clarified myths and misconceptions about the disease. The positive effects of lifestyle changes were mentioned in less than 25 percent of the ads and fewer than three out of 10 acknowledged that older treatments were available. Two out of five ads attempted to medicalize ordinary life issues. (Routine hair loss or a runny nose, for example, became a medical problem requiring treatment with expensive prescription drugs.)[5]

Contributing to the effectiveness of these drug ads are certain mistaken beliefs held by the public. Half of all respondents to a survey thought each drug ad was approved by the government before it was shown to the public and 43% thought that the drugs permitted to be advertised were exclusively those that were entirely safe.[6] These erroneous beliefs held by people exposed to drug ads add to the urgency and necessity of tighter regulation of drug ads by the FDA: the public is genuinely being harmfully deceived.

To begin with, the FDA should ban ads that are engaged in disease mongering, and heavily fine any company that does not immediately comply to pull such an ad when ordered to do so. These are ads that characterize normal, ordinary experiences and conditions as illnesses that people need not suffer from if they take the drug named in the ad. By means of such ads, drug manufacturers are mercilessly inventing diseases to expand markets for their drugs.

In Chap. 6, we illustrated just how exceedingly harmful disease mongering can be. I will here remind the reader of ads for Paxil, an antidepressant. In the psychiatric manual (DSM-IV) social anxiety disorder (SAD) and generalized anxiety disorder (GAD) occur when an individual has multiple, persistent symptoms associated with severe distress and impaired functioning. Ads for Paxil promise relief for being "allergic to people" and help for SAD sufferers to brave dinner parties and public speaking. By consulting the DSM-IV, the FDA can identify false accounts of illnesses in such drug ads and have the ads stopped and heavily fined if need be. Disease mongering, as pointed out in Chap. 6, contributes to the considerable avoidable harms of overdiagnosing, overtreating and overmedicalizing individuals. Antidepressants, inappropriately prescribed, can be fatal. Recall that they lead to excessive deaths in nursing homes and make adolescents, so prone to committing suicide and having violent impulses, that antidepressants are not recommended for being prescribed for adolescents in the United Kingdom.

---

[5] Abramson, Overdosed America, 154.
[6] Ibid, 154–155.

The FDA has to insist that the labels and ads describing drugs plainly and completely list all of the drug's side effects. To do less than that, deceives viewers and readers of ads for drugs into thinking a given drug is safe. As noted above, 43% of respondents to a survey thought the drugs being advertised were completely safe. This same survey found that 50% of those surveyed thought that these ads were approved by the government prior to their release. These people would be quite shocked to learn that the FDA actually allowed Vioxx for too long to omit references to its very harmful side effects, in particular causing an excess of heart attacks, and later allowed that side effect to appear where it would not likely be noticed. And its most serious harmfulness was not generally noticed. Vioxx brought in billions in revenue before a study was published that revealed just how harmful the drug really was, something the FDA knew about before that study. Indeed, drugs like Vioxx, as their own scientists warned, should not be approved by the FDA in the first place!

7.1.1.3 Refrain from Practices that Facilitate or Protect the High Prices of Drugs

If the FDA were to follow all of the suggestions enumerated so far, expenditures for high priced drugs would be considerably reduced. Many of the drugs now approved would not have been because they would lose out to existing equally or more efficacious, cheaper drugs or to therapies very safe and relatively much less expensive like diet and exercise. Thus, for example, billions would not have been spent for Vioxx had the FDA simply gone with the scientific knowledge it possessed: the FDA knew that Vioxx was no more efficacious and less safe than the much cheaper over-the-counter Aleve; there was no need for Vioxx nor a justification for such a harmful drug. As we have already noted also, diet and exercise and dropping bad habits like smoking can for some conditions outperform drugs, the approval of which is thereby rendered unnecessary.[7]

We remind the reader that the FDA's mission includes making medicines more affordable. In light of that mission, the FDA should not be in the business of protecting high prices charged for drugs in the U.S. The FDA should not persist in discouraging and effectively banning the importation of identical, lower priced FDA approved drugs. Such drugs are much cheaper in Europe and Canada.[8] In 2001, without providing evidence, the FDA falsely testified in Congress that these drugs are not safe.[9] Furthermore, the FDA publically asserts that it is illegal to import

---

[7] See ibid., 245–247, for examples, of high priced drugs unnecessarily approved because they are no more efficacious and sometimes less safe than much less costly drugs approved and in use earlier. In some instances there are interventions that are much more efficacious, safe, and inexpensive than any of the drugs in use.

[8] For a sample list of such drugs see William Faloon, Pharmocracy: How Corrupt Deals and Misguided Regulations Are Bankrupting American – and What to Do About It (Mount Jackson, VA: Praktikos Books, 2011), 280.

[9] Ibid, 276–278.

these drugs and threatens importers with dire legal consequences.[10] At the same time, the FDA permits U.S. drug companies to import active ingredients they have had synthesized in other countries the FDA says cannot be trusted. These ingredients are in many U.S. drugs: this suppression of competition adds to the profits made from these highly priced drugs, prices that exceed those charged in all other countries. Congress should assert its authority over the FDA and stop the FDA from blocking much less costly FDA approved drugs from other countries. Billions of dollars would be saved annually, thereby greatly reducing costs for health insurance companies, assisting medicare and Medicaid achieve solvency, and significantly easing the current financial burdens facing the public, both as patients and taxpayers.

There is another avenue for lowering drug prices. The prices for generic drugs could be substantially brought down if the FDA were to change how the manufacture of generic drugs is regulated. At present, any company that wishes to manufacture a generic drug, whether as a prescribed drug, or as one sold over the counter, has to file an Abbreviated New Drug Application (ANDA) with the FDA even if others make it. If proof of bioequivalence is required, the company will need to undertake human studies for 1.5 to 2 years unless such studies exist. Still, the FDA can reject the ANDA and require the company to do new studies.

This approval process usually costs around two million dollars and extends the time to get the drug to market to two to three years. Over-the-counter drugs do not require a company to perform its own research, and that reduces the cost to about one million dollars and the time for marketing it to 1.5 years. Companies also need to set aside 15% of their budgets for legal fees, since any large manufacturer with a sizable existing market share, will be prone to sue in an effort to ward off competition.[11]

The physician Willian Faloon has suggested that this whole process of having ANDAs unnecessarily drives up the cost of generic drugs. At a much lower cost, there are manufacturers that could put the ingredients of these drugs into pills using the same processes they employ to produce pills or capsules for vitamins and other nutritional supplements.

Faloon illustrates how this would work using Finasteride (Proscar) as an example.[12] This drug helps relieve benign prostate enlargement and may also reduce the risk of cancer. Finasteride came off patent in 2006, and in 2008, chain pharmacies were charging as much as 90 dollars (86 on average) for a 1 month supply of thirty tablets. This drug simply puts five mgs Finasteride into a tablet that dissolves in the stomach. Vitamin companies do this with nutrients but the FDA does not allow them to do this with drugs.

Given the cost of Finasteride and inserting it into tablets, thirty tablets could be profitably sold for 10 dollars and 25 cents. Taking into account the considerable benefits of Finasteride, Medicare would save a great deal of money if the FDA

---

[10] Ibid, 280–287.
[11] Ibid, 138–139.
[12] Ibid, 137–138.

would no longer prevent Good Manufacturing Practice (GMP) certified vitamin producers from manufacturing and marketing drugs. These savings would apply to quite a number of drugs that Americans in 2009 were paying 837% more to purchase at chain pharmacies than what they would pay for these same medications manufactured and sold by vitamin and supplement producing companies.[13] One should note also that drug prices exacted by drug companies keep rising, quite sharply at times.[14]

Having drugs made by CMP-certified vitamin companies is safe. The FDA regulates and inspects the facilities of these manufacturers and tests samples of their products. Any failure to meet FDA specifications would have the prospect of being confronted with governmentally assessed civil and criminal penalties.[15]

A reformed FDA can further help patients obtain much cheaper drugs from compounding pharmacies. As things now stand, these pharmacies can only advertise their existence. However, as Faloon indicates, by law they cannot publish the name of a drug he refers to for illustrating the enormous money to be saved by purchasing it from a compounding pharmacy. The price of this drug made by a large pharmaceutical company is 245 dollars a month; at a compounding pharmacy, the same drug costs 29 dollars a month, saving the purchaser 2592 dollars a year! There should be no legal restriction on publishing information that names the drugs and the prices attached to them at compounding pharmacies.

Still another failure to promote affordability involves fish oil. The FDA has approved fish oil as a prescription but not as a natural supplement.[16] The labels of natural supplements of fish oil have to state that the fish oil has not been evaluated by the FDA and is not intended to diagnose, treat, care or prevent any disease. At the same time, manufactured as a prescription by drug companies can state the health benefits of fish oil. In yet another instance, expensive drugs are promoted while relatively inexpensive, natural substances equally efficacious, and in some instances more efficacious, are not recognized as such by the FDA; another reason the FDA must be reformed.

7.1.1.4 Assuring that New Scientific Research Is Accurate and Made Available to the Public

The FDA has these responsibilities specifically included in their mission statement. That was documented in Chap. 6 as well as extensive existence of deceptively conducted and deceptively reported research by drug companies. The remedy for one of

---

[13] See a list of current generic drugs, comparing the average prices charged at chain pharmacies with what free market prices would be if, for example, produced by vitamin manufacturers in ibid., 140–141.
[14] See a sample list in ibid., 143 reprinted form the April 15, 2009 issue of the Wall Street Journal.
[15] Faloon, Pharmacracy, 144. See also 145–147 for some of the benefits of less expensive drugs.
[16] Ibid, 203–206.

the very serious distortions of scientific data is one that has twice been addressed by Congress but has never been carried out nor enforced by the FDA.

Then the FDA Amendment Act was enacted in 2007, tightening and widening requirements for registering to include all drug trials and those of medical devices being developed and for which approval to market is being sought. Furthermore, the result of trials are to be posted after 2007 in abbreviated summary tables within the year they were to be completed. That led people to think the deleterious problem of missing data had been eradicated. That hasn't happened. For one thing, the Act does nothing for shedding light on the reliability of many drugs and medical devices currently being prescribed. What is more, these regulations are largely ignored. A study of first trials subject to mandatory reporting discovered that no more than one in five had posted their results. A 10 thousand dollar a day fine did not sufficiently work: three and a half million dollars a year is nothing for any drug company to pay that has annual sales measured in billions of dollars. Besides, no fine has ever been levied.[17]

Once again we find that the FDA fails to carry out its duties and, in this case, does so in spite of express legislation backing for what these are. Of course the fines for noncompliance with the FDA Amendment Act of 2007 should be much greater and those in non-compliance should not be allowed to have their products marketed, whether drugs or medical devices.

Demanding and enforcing the pre-registration and posting of the trials would enable the FDA to check the studies for their scientific integrity. The list of the various ways in which research has been rigged and deceptively presented by drug and medical device manufacturers for portraying their products in a favorable light is a long one indeed: the documentation of all that can be found in Chap. 6.[18]

Given the FDA mission to help make the public informed by accurate scientific knowledge, its suppression of the scientific studies being cited by those marketing sour cherries and walnuts, as documented in Chap. 6, cannot be justified. That is particularly true since the FDA did not and could not dispute the scientific findings they were not allowing to be freely made available to consumers, findings that commend these foods as nutritionally beneficial, contributing to being healthy and well. The FDA should surely also approve such inexpensive and safe natural means for preventing and curing illnesses whenever the science supports doing so.

7.1.1.5 Dealing with the Deleterious Influence of Conflicts of Interest Within the FDA

What should be abundantly clear is that the FDA is falling far short of fulfilling its mission. Remedying that situation would go a long way to putting medical care on the high moral ground, the achievement of which, would significantly enhance the safety and efficacy of warranted medical interventions, and greatly reduce the resort to medical interventions that are not warranted. The result would be considerable

---

[17] Ibid, 52–54. Note that this was published in 2012.
[18] Ibid, 176–222 provides documentation of what Goldacre calls "bad trials."

reductions in expenditures. The flawed and unjustifiable practices and policies are not as a rule due to a lack of knowledge necessary to succeed in carrying out its mission. How then can we account for the FDA's many failed practices and policies?

The reader will recall that the FDA not only approved Vioxx knowing its harmful side effects but also voted not to withdraw it when further evidence of its harmfulness came to light. I remind the reader that after Merck withdrew Vioxx, David Graham, the Associate Director for Science and Medicine in the Office of Drug Safety within the FDA testified before the U.S. Senate Committee on Finance. In his testimony reporting the serious harms caused by the use of Vioxx, Graham had this explanation for the disastrous FDA decisions with regard to Vioxx:

> The FDA has became an agent of industry. I have been to many, many internal meetings, and as soon as a company says it is not going to do something, the FDA backs down. The way it talks about "industry" is "our colleagues in industry."[19]

Astonishingly, the FDA is being described as an agent of industry. I use the word astonishingly quite advisably. After all, the FDA is officially an agent of the U.S. Federal government, not an agent of industry.

Still, those who manufacture medicines and medical devices are among the sources expected to advance the efficacy of medical care. Hence, the FDA does have a positive interest in what drug and medical device companies can successfully offer on behalf of improving the health of the public and thereby assist the FDA to accomplish its mission. At the same time, the FDA is an agent of government to judge whether any product alleged to be one that will improve the health and well being of the public is likely to do so. To make that judgment, the FDA assesses whether a given product can be scientifically demonstrated to be efficacious, safe, and either superior to products in use or genuinely therapeutically totally innovative. That is how the FDA is expected to act. How then does it act when it can appropriately be called an agent of industry, and for that reason, failing to carry out its mission to enhance the health of the public?

There is no question about whether some drugs, medical devices and technology improve health, prevent diseases, reduce suffering, and save lives. Surely it is plausible to see this as the primary aim of the companies that manufacture these products. Unfortunately, that is not the whole story. As private companies they have another aim: it is to be as profitable as possible, and attract and satisfy investors to assure success. What is unfortunate about this aim is that the quest for profits is so intense that, far too often, the motivation to assure the health and wellbeing of the public is undermined or overwhelmed by this fervent quest for profitability. We saw that happen in the Vioxx case. In that case and cases like it, the FDA's assigned task of securing and improving the health of the public is similarly derailed by acting on behalf of Merck's profitability by approving an unneeded product and deciding to keep Vioxx bringing in profits despite its known disastrous side effects on the health of its users. It is for facilitating industry's strong interest in profitability that leads

---

[19] Goldacre, Bad Pharma, 127–128.

Graham, their own scientist, to regard the FDA as an agent of industry. What follows now is an example of how that takes place within the inner workings of the FDA.

In 1996, Warner-Lambert sought approval from the FDA for its diabetic drug Rezulin.[20] Dr. John L. Geriguian, a 19-year veteran of the FDA, was the medical officer who evaluated the application on behalf of the drug Rezulin. He recommended that Rezulin not be approved. The drug was no better than drugs already being marketed and it had a troubling tendency to cause inflammation of the liver. Executives at Warner-Lambert approached the higher-ups at the FDA with a complaint about Geriguian. Subsequently, Dr. Geriguian was removed from any participation in the approval process of Rezulin. At the meeting at which the Advisory Committee was to decide whether to approve Rizulin, they were not told about the liver toxicity that led to Dr. Guerigian's view that the drug should not be approved. In February, 1997, the FDA approved Rezulin. It sold briskly; it was regarded as a "blockbuster."

What Dr. Gueriguian feared happened. Soon reports of deaths due to Rizulin's fatal liver toxicity surfaced in the U.S. and Japan, and because of its liver toxicity, the United Kingdom withdrew the drug in December 1997. Despite all of these developments and the ever increasing liver problems in the U.S. induced by taking Rizulin, the FDA did not withdraw it from the market until March 2000! That delay allowed Warner-Lambert to garner 1.8 billion dollars from the sale of Rizulin and that delay allowed a suspected 391 deaths and 400 cases of liver failure to occur.

The FDA is indeed an agent of industry. The FDA, in too many instances, helps to assure that the products of companies they assess are profitable even when they know that these products are not safe and efficacious enough to be marketed. As their own scientist Gueriguian told the Los Angeles Times: "Either you play games or you're going to be put off limits—a pariah." The Los Angeles Times told the story of Rizulin in 2010: It went unheeded.[21]

The FDA's flawed approval of drugs has a long history. A survey by Public Citizen surveyed medical officers of the FDA, and those who responded cited 27 new drugs approved in the three years pervious to the survey that they believed should not have been approved. A report by the inspector general of the U.S. Department of Health and Human Services, published in March 2003, indicated that one-third of the medical officers in the FDA's Center for Drug Evaluation and Research professed to being uncomfortable about voicing their differing opinions. And these officers had good reasons for being uncomfortable. Again in the Los Angeles Times, it was reported that seven drugs the FDA had approved since 1993 had to be taken off the market since they caused such serious side effects, among them more than 1000 deaths. Since reporting adverse side effects is voluntary the number of deaths could be a great deal more than a 1000. As the Los Angeles Times reported, "the FDA approved these drugs while disregarding danger signs or blunt warnings from its own specialists."[22] Additionally, just imagine the fact that there

---

[20] Abramson, Overdosed America, 87–88.
[21] Ibid, 86–87.
[22] Ibid. See 85–90 for a fuller account of FDA's flawed approval process.

were one of every ten adults, 22 million people in the U.S. who had used a drug between 1997 and 2000 that was later withdrawn.

There you have it: those in position to set policy in the FDA are without excuse. They have the scientifically qualified personnel, the power, and the mandate to avoid the scientifically and ethically unjustified decisions and decision-making processes that cause so much suffering and so many deaths as we have illustrated and documented in Chap. 6 and so far in this chapter. And, in the bargain, the FDA is wasting billions of dollars annually. At this juncture we must address what is driving such marked deviations from ordinary empathy for those who would, and those who do, suffer and die from medical interventions that the FDA could and should have prevented or quickly halted but did not do so. Unquestionably, the decisions of the FDA are influenced by substantial conflicts of interest. This situation should obviously be corrected.

To begin with, sociologists who study regulation have observed what they call "regulatory capture." The concept refers to a process that succeeds in having government regulators act on behalf of the industry they have the responsibility to monitor rather than the public interests they should be serving. How this is carried out has been thoroughly documented in the scholarly literature: Goldacre quotes from that literature this key passage:

> Effective lobbying requires close personal contact between the lobbyists and government officials. The object is to establish long-term personal relationships transcending any particular issue. Company and industry officials must be 'people' to the agency decision-makers, not just organizational functionaries… Officials will be much less willing to hurt long-time acquaintances than corporations. Of course there are also tactical elements of lobbying… This is most effectively done by identifying the leading experts in each relevant field and hiring them as consultants, or giving them research grants and the like. This activity must not be too blatant, for the experts themselves must not recognize that they have lost their objectivity and freedom of action.[23]

We know that regulatory capture works: it has been very successful in capturing the FDA officials. Officials at the FDA refer to their counterparts in the medical industries as "their colleagues." Studies show that physicians think that they themselves are not influenced by their contacts and connections with drug companies even though physicians generally are. The FDA knowingly draws on experts with financial ties to drug companies to serve on the Advisory Committees that make recommendations as to whether to approve or disapprove a particular drug. At the meetings held for that purpose, federal law generally prohibits such a role for experts with financial ties to the product being considered. However, the FDA grants many waivers to such experts. For example, between 1998 and 2000 it was 800. Consequently, 54% of the experts on these vital Advisory Committees served with a direct financial interest in the drug or topic they were recruited to assess.[24] In addition, experts who served did not need to declare less than 50 thousand dollars

---

[23] Goldacre, Bad Pharma, 123–124.
[24] Abramson, Overdosed America, 89–90.

per year from a drug company, provided the payment is not for work related to the drug under discussion.

A reformed FDA should adopt and abide by much more stringent rules governing the membership of their Advisory Committees. Those who serve should have no financial ties to the company whose product they would be evaluating for possible FDA approval. No financial ties means no money or monetary gain of any kind from that company. Experts are bound to have some financial relationship with some drug or medical device companies such as support for their research. However, they should not serve on Advisory Committees of the FDA judging the merit of the products of those companies.

The present FDA officials are so entangled in positive relationships with officials in the medical industries that they exhibit all the symptoms of "regulatory capture." New personnel, without any conflicts of interest we have described, should be carefully selected to reform the FDA and rescue it from engaging in activities that cannot be ethically, scientifically, and financially justified.

A significant conflict of interest perceived by many as compromising the work of the FDA, stems from its substantial financial dependence on the very industries seeking FDA approval for their products. These manufacturers pay the FDA costly fees to have their products evaluated, providing something on the order of 50% of the FDA's budget. To remove this conflict of interest, the FDA should be completely funded by the federal government. If the FDA were to carry out its mission properly and impartially, it would save hundreds of billions of dollars annually by no longer approving so many non-efficacious and harmful medical products and intervention, by allowing competition from imported drugs, and by giving approval to scientifically supported nutritional sources of preventing and curing illnesses, something the FDA currently will not do.[25]

What then should be done about fees? They should not be abolished. What the FDA does should be seen as a vital public service taxpayers provide. Charging a fee for this service from those who stand to benefit from its use is appropriate. However, the current fees should be lowered, perhaps generally, but at least for those who do not have the resources of the pharmaceutical giants. The money received from these fees should not go to the FDA but rather to the federal government's budget for Medicare and Medicaid to assist in providing the full coverage of medical care necessary for achieving justice. With this policy in place, it would be one of the ways those who manufacture medicines and medical devices and technology would be sharing their resources with those who need medical care, a hallmark of the traditional professional ethic of physicians since at least the time of Hippocrates.

To remove a further conflict of interest, employees of the FDA should not be making decisions while contemplating a future as employees of pharmaceutical, medical device or medical technology manufacturers. Any future employment with

---

[25] Ibid, 249–253 and Hadler, Citizen Patient, 178–191. Abramson includes many of the suggestions I have made that should guide FDA decisions. Hadler advocates a replacement for the FDA and the reform of it much as I have outlined it.

any of those companies should only be possible several years after leaving the FDA. Congress should make a judgment about the length of that interval.

Making sound objective judgments that are scientifically and ethically justifiable demands of those making them a particular orientation and set of criteria. What this entails is a commitment to make morally justifiable decisions, making use of cognitive processes of the sort identified in Chap. 2. First of all learn as many facts as one can relevant to making a decision at hand. Secondly, one should vividly imagine how people will be affected by a decision being contemplated. Lastly, on has to make any decision in question from an impartial perspective. Recall that one cannot put a limit on the facts one needs or on the people that may be affected to assure that a decision will be right or best one to make. That means that any decision one makes is subject to possibly needing future corrections, alteration, or even outright rejection: therefore decisions, once implemented, should be carefully monitored.

I think it is important to examine what achieving an impartial perspective demands of FDA officials. How, for example, is impartiality to be sought when trying to decide whether to approve the marketing of a particular drug? On the one hand, the FDA has an obligation to foster innovation for the sake of improving public health; on the other hand, it exists to protect the public from any harmful products and to prevent costly, wasted expenditures on products that are no better than, or inferior to, products already in use for a similar purpose. To achieve impartiality is to be as conscientiously empathic toward the public and its health as one is toward those who allege to have a product that will, in a new way, or to a higher extent improve the health of the public. Adopting such a perspective is to overcome as nearly as possible the conflicting or competing interests confronting FDA officials in the approval process: to obtain objectivity requires attaining an impartial perspective.

When following its mandate, the FDA is pursuing the demands of justice, particularly avoidance of harm, the relief of suffering, and the duties to heal and rescue. Ideally, those who manufacture drugs, and medical devices and technology are pursuing these same moral demands. In any event, the FDA's mission is to see to it that the products and foods they approve meet these demands of justice. All of the suggestions for reform of the FDA I have identified are intended to satisfy these imperatives of justice.

7.1.2 Additional Governmental Policies for Reducing Excessively High Prices of Drugs, Medical Devices, and Hospital Services

Reforming the FDA, as already suggested above, should help reduce prices and many other costs associated with the provision of health care. However, more can and should be done to assure the sustainability of meeting everyone's genuine need for medical interventions.

Although drugs imported from Europe would likely reduce drug prices in the U.S. somewhat, the Congress should nevertheless amend its policy enshrined in the Affordable Care Act (ACA) and negotiate the prices of drugs. As we have reported, the cost of drug ingredients is very low and the sale of drugs in Europe is more than sufficiently profitable. In countries like India, prices are much lower. We have documented that in Chap. 6.

The prices of medical devices should also be negotiated. That requires Congressional action as well. Take one example of why this is necessary. Medronics manufactures spinal products and therapies that cost 24.9% of what they sell for, thereby yielding the extraordinary gross profit of 75.1%. Apple's robust profit margin for its high tech products is 40%.[26]

The ACA makes a modest effort to recover some of the cost of the exorbitantly priced medical devices: it levies a 2.39% [excise] tax on the gross profits of their manufacturers. The problem with this approach is that it does not reduce prices. In fact, the companies making medical devices are lobbying Congress to remove this tax, arguing that it forces them to raise prices and suffer reduced sales and a reduced workforce as a result. It is certainly not the case that the tax forces those who produce medical devices to raise their profits: the 2.39% excise tax only reduces their gross profits to 50%, a 10% higher profit than that of Apple.[27] It would probably be a good idea to have in place a tax on profits and also negotiate prices. However, a tax on profits should be on operating profits, that is, on the excess of revenue over expenses. That would facilitate and encourage new start-up companies and help small companies survive temporary economic downturns.

The urgency of reducing prices applies as well to the squeeze put upon hospitals who currently have to cope with what they are charged for the medications, medical devices, and for some of the other supplies needed to provide proper care for their patients. On average, full service hospitals have low profit margins of 1–3%: one half didn't break even in 2009; and from 2010 until 2016, 80 rural hospitals have closed.[28]

Yet, at the same time, there are hospitals that are exceedingly profitable.[29] Recall from Chap. 6 that the cancer center A D Anderson bought a very expensive cancer drug for $3000 per dose, and then turned around and charged a patient $13,702 for one dose. As we have documented, the drug company could have made a handsome profit at a much lower price and obviously so could A D Anderson. Prices at such levels contribute mightily to the unsustainability of the U.S. healthcare system.

Children's hospitals are also highly profitable. In 2009, for example, Texas Children's Hospital achieved a profit of 275 million dollars and Children's Hospital of Philadelphia (CHOP) a profit of 359 million dollars. On top of that, CHOP

---

[26] Stephen Brill, "Bitter Pill: How Outrageous Prices and Egregious Profits Are Destroying Our Health Care," Time (March 4, 2013), 36.

[27] Ibid, 34.

[28] Finding of the University of North Carolina Health Research Program.

[29] Brill, Bitter Pill, 31 has listed 10 of the largest, most profitable non-profit hospitals. Recall that the top one achieved an operating budget of $769,706,094 and paid its CEO $5,975,462.

received a government grant worth 121 million dollars.[30] Being subsidized by the government is not enough for children's hospitals: they raise about one quarter of their revenue from donations.[31] Consider that Children's Hospital of Boston made 111 million dollars in profits, paid millions to its CEO, and yet asked children to collect pennies from other children at school.[32] Indeed, their fund-raising is so lucrative that Children's Hospital Boston employs no less than 125 full-time fund-raisers. No doubt donors think that their money goes directly to caring for sick children. I wonder what would happen if donors knew how much is spent on salaries for CEOs. In the same year (2009) at Children's Mercy Hospital in Kansas bombarded their community with fund-raisers, its CEO was paid $5,987,194.[33] That same year, the CEO of Children's Hospital of Wisconsin, Milwaukee received $5,465,948 in compensation, and the CEO of Children's Hospital Medical Center of Akron, Ohio received $5,132,104.[34] These are all non-profit hospitals!

One thing is clear. Highly profitable non-profit hospitals should not be receiving government subsidies. Congress should determine that non-profits above a certain level should not be eligible for government subsidies. Furthermore, Congress should follow the same policy it has for health insurance companies: the percentage spent for administrative costs should be limited.

As a guide as to when to withdraw subsidies from non-profit hospitals, Congress might consider the policy adopted by Martin Makary, a surgeon at Johns Hopkins Hospital and an associate professor of health policy at Johns Hopkins School of Public health. He donates money for charities that fight child disorders. Appalled by the excessive profits of children's hospitals and the salaries of their CEOs, he only gives money to such charities if their annual surpluses amount to less than 100 million dollars annually.[35]

Withdrawing subsidies of highly profitable non-profits hospitals still does not directly affect those non-profits who have operating profits that top 100 million dollars annually. Congress could tax those profits. To justify taxing hospitals for all hospitals with surpluses over a specific amount, requires that the money raised in this way be used for assuring that patients at those hospitals receive the medical care they need regardless of whether they can afford it; and that includes patients whose insurance is insufficient to cover all the costs they incur.

What the policies of taxing the operating profits over a certain level being garnered by drug companies, medical device and technology companies, and hospitals are actually instituting, is the very kind of profit sharing that is the hallmark of the traditional medical ethic. Physicians loyal to the Hippocratic ethic used money

---

[30] Marty Makary, Unaccountable: What Hospitals Won't Tell You and How Transparency Can Revolutionize Health Care (New York: Bloomsbury Press, 2012), 132.

[31] Ibid, 134.

[32] Ibid, 131.

[33] Ibid, 130–131.

[34] Ibid, 130.

[35] Brill, Bitter Pill, 54–55. Brill has also suggested taxing hospitals including salaries of CEOs. Readers may wish to consult a number of his suggestions and his rationale for them.

earned from their fees to offer medical care to those who could not pay fully or nothing to receive it.

Recall that the American Medical Association promulgated that Hippocratic ethical ideal but dropped it in 1980. The prices then and now make it extremely difficult for individual physicians to provide care free or at reduced prices, though some make every effort to do so to the extent that they can. To make the Hippocratic ethic work without government intervention, voluntarily in the present circumstances, requires that all companies, organizations, and individuals that contribute in any way to the medical care of patients be committed to practicing it.

Businesses do not have to be rapacious to be profitable. Market Basket owns and operates a considerable number of grocery stores. Market Basket has a policy of sharing its profits with employees, and, with lower prices than its competitors, in effect shares its profits with its customers. Yet it is profitable and has been adding new stores.[36] The companies that make drugs, and medical devices and technology, could be profitable following the practices of Market Basket. Indeed, as those who are serving patients who desperately need their products to avoid suffering and death, one would think they would be prompted by sheer mercy to be sure that these patients can afford to receive the care they need. Such expressions of mercy shaped the time-honored ethic articulated by the Hippocratics and by the parable of the Good Samaritan.

7.1.3 Government Interventions and the U.S. Health Care System as a Market

Some of the governmental interventions I have been suggesting introduce more competition into the health care market: others introduce some new restraints into that market. It is important to see why both kinds of Congressional measures are necessary given the inherent uniqueness of the healthcare market.

Without the kinds of interventions I have been suggesting, the present health care system does not, at least for the most part, function as a free market, the investigative reporter Stephen Brill came to the conclusion that the U.S. health care system does not function as a free market at all based on his research of healthcare costs and the unfairly and unnecessarily high prices of drugs, medical devices and equipment, and exorbitant charges for services. Brill has this to say on behalf of government:

> Put simply, the bills tell us that this is not interfering in a free market. It's forcing the reality that our largest consumer product by far—one fifth of our economy—does not operate in a free market.[37]

---

[36] For a brief description of the successful business model of Market Basket, see Dave Solomon, "Market Basket Protest: A Year Later," New Hampshire Sunday News, July 5, 2015, vol. 69, No 39, pages A1 and A4.

[37] Brill, Bitter Pill, 54.

Brill advocated his own version of taxes on profits and some other policies that readers would do well to consult.

Brill does not address the role of the FDA in suppressing competition and what congressional reform of the FDA can do to facilitate more competition in the health care market. Some of the suggestions I have offered on behalf of reforming the FDA would, if adopted and implemented, free up competitive forces in the health care system and greatly reduce health care costs. I remind the reader of the following suggestions: allowing importation of lower priced, FDA approved bioidentical drugs from Europe; allowing public access to the prices of drugs from compounding pharmacies; reducing the cost of generic drugs further by having vitamin manufacturers make the pills, reducing application fees for seeking approval of new medications, medical devices, and medical technology; and approving on scientific grounds, naturally nutritious foods and nutritional supplements derived from such foods, superior in efficacy and safety to existing drugs for preventing and/or curing illnesses.

Some might argue that creating a more competitive health care market is the way to reduce prices. I would argue that a more competitive health care market is a way to reduce prices and it certainly should be facilitated. However, health care is an especially vital service. Those who cannot obtain it when they genuinely need it are faced with the prospects of suffering from debilitating illnesses or even death. Understandably then, people with the need for it, do not regard health care as a dispensable purchase. It is in these kinds of circumstances that the government is moved to intervene because these are the kinds of situations that invite price gouging. We illustrated that phenomenon in Chap. 6; cancer drugs sought by individuals who fear dying was one of the prime examples.

While this chapter was in progress, the U.S. Congress and Justice Department went after certain U.S. airlines for what they regarded as price gouging. When a passenger train derailed in the vicinity of Philadelphia, it left quite a number of people without transportation they very much depended upon for some weeks. Airlines allegedly raised their prices for traveling the route in question and the government immediately responded. Unfairness of this kind is not regarded as legally permissible.

Consider also that housing and groceries are in competitive markets. Nevertheless, they too involve products that people require. In those cases, the U.S. government does have policies to encourage or provide low cost housing and shelters, and food stamps for those who cannot afford to buy the food necessary for meeting their nutritional requirements sufficiently. There is government assistance for providing food for needy children in the public schools as well.

My response to price gouging is to tax profits that are excessive enough to qualify as price gouging. Doing that generates income that can and should be shared with patients who are uninsured, or inadequately insured, or otherwise unable to pay for the medical care they need. I have no problem with Congress and the Justice Department resort to fines for price gouging. I would urge, however, that the money raised be used to treat patients who are unable to pay sufficiently or at all for their necessary medical care. From such an approach, everyone benefits in the long term.

The whole health care system, in fact the entire U.S. economy cannot sustain the perpetuation of the present excessively rapacious pursuit of profits. With the enormous amount of money currently gleaned in the health care system, sharing these resources is essential to avert its financial collapse.

Let me emphasize once more that the Hippocratic ethics of sharing one's profits with patients should not be regarded as a naïve relic of the past: it is the key lynchpin for attaining and perpetuating a health care system that sustainably meets everyone's needs for medical care; it is no less a key lynchpin for attaining and perpetuating human communities. Mercy (justice) is a requisite of community; no community can exist or last if the practice of mercy does not permeate every one of its sectors and organizations, and if its members are not instructed to recognize mercy's indispensability for their communities, their health and well being, and their very lives. I remind the reader of Robert Frost's recognition of mercy's indispensability: "Nothing can make injustice just but mercy."[38]

7.1.4 Rethinking Patenting in the Medical System

"There was a time not so long ago when breakthroughs in medical science were driven more by health needs than by the search for corporate profits."[39] That is how Abramson begins the concluding chapter of his book arguing his case for saving at least 500 billion dollars annually while improving the health of Americans. He offers an example of an outstanding breakthrough in medical science that was done to meet health needs, not profits: it was Dr. Jonas Salk's development of the polio vaccine. Amid all the publicity surrounding the vaccine's release, Dr. Salk was asked who held the patent. To that he replied: "Well the people, I would say. Could you patent the sun?"[40] How refreshing! Like a cool ocean breeze on a hot day, Dr. Salk's quest was fueled by sheer compassion for all those who would suffer from polio without his vaccine, some of whom would have died. His quest was clearly not personal enrichment; rather it was to make the vaccine as freely and cheaply available as possible to everyone.

Among our suggestions for reforming the FDA, was that of properly reducing the approval of relatively worthless and seriously harmful drugs. That would cut down considerably on the number of high priced, patented drugs in the market. Still, there is more that can be done to make the medical market more competitive. One possibility is to cut back on the amount of time patents can remain in effect; another is to set a higher bar for the inventiveness required before a patent is granted. In any event, the rationale for issuing patents for products that can prevent or treat illnesses

---

[38] Robert Frost, "A Masque of Mercy." In the Poetry of Robert Frost: the Collected Poems, Complete and Unabridged, Edward Connery Latham, ed. (New York: Holt, Rinehart and Winston, 1979), 521.
[39] Abramson Overdosed America, 241.
[40] Ibid.

and disabilities should be addressed by Congress. Patents in the U.S. health care system are creating monopolies. And, we do after all generally oppose monopolies and the price gouging usually associated with them.

7.1.5 Tort Reform

Currently, there are a number of very costly situations that cry out for tort reform. First, there is widespread resort to defensive medicine. The threat of being sued does lead to administering many more highly expensive tests than are necessary. One hospital CEO told the investigative reporter Stephen Brill that:

> We use the CT scan because it's a great defense. For example if anyone has fallen or done anything around their head—hell, if they even say the word head—we do it to be safe. We cannot be sued for doing too much.[41]

The costs of such excessive, uncalled for, uses of CT scans can be very considerable indeed. Take the case of Emilia Gilbert who fell on her face in her back yard and broke her nose. She went to the emergency room of Bridgeport Hospital, owned by the Yale Haven Health System. Gilbert received three CT scans of her head, chest and face for which she was billed $6538. Medicare would have paid approximately $825 for all three. Gilbert did not have Medicare. On top of that outrageous bill, she also received high doses of radiation that unnecessarily exposed her to the risk of getting cancer.

In Gilbert Welch's book discussing the problems of overdiagnosis, he takes note of the role played by defensive medicine, a practice he refers to as "safety first."[42] Though the perceived risk of malpractice is greater than the actual risk, it is the perception that matters. After all, there are legal penalties for failure to diagnose and not for overdiagnosis. What's "safer" is obvious. Welch claims that most doctors, if asked to choose the most powerful incentive for undertaking more diagnoses, would most likely choose the one that involves lawyers. Welch refers his readers to a study that found that "a doctor's being sued for failure to diagnose prostate cancer leads not only the defendant to order more PSAs but also his or her colleagues."[43]

However, as Welch goes on to observe, here is no evidence that screening for many conditions will reduce an individual's chance of developing an advanced disease. Even in instances where we do know that screening helps some people, others with rapidly growing cancers are not helped because the cancer was not discernable at the time of screening or was simply resistant to treatment. Thus, for example, PSAs and even better tests will not prevent those screened from dying from cancer. Yet one of Welch's colleagues was found guilty of malpractice for failing to order a

---

[41] Brill, Bitter Pill, 24.
[42] H. Gilbert Welch, Lisa M. Schwartz, and Steven Wotoshin, Overdiagnosed: Making People Sick in the Pursuit of Health (Boston, MA: Beacon Press Books, 2011), 161.
[43] Ibid, 163.

PSA screening even though he had found no cancer on the basis of a digital rectal exam. Subsequently his resort to PSA screening increased significantly.

The Choosing Wisely Movement carried on by profession medical organizations is in the process of issuing numerous recommendations to abstain from resorting to tests and screening devices they deem unnecessary, only harmful, and a waste of money. Compliance with such recommendations would be more predictable and welcome if tort reform were to make compliance safe.

The costs of defensive medicine are difficult to ascertain. One study focused on the cost of malpractice insurance to try to determine how much money could be saved by tort reform.[44] The cost of malpractice insurance amounts to less than 2% of national health care expenditures The study claimed that tort reform would reduce insurance premiums by 10%. Using that measure they found that what was saved would constitute less than 1% of medical care costs. Still the authors regard tort reform as worthwhile.

The study above does not attend to the largest cost of defensive medicine, namely the cost of procedures being used. One study reports an estimate of 45.6 billion dollars for hospital and physician costs, and a total of 55.6 billion dollars when all costs associated with the liability system are taken into account: this is for the year 2008.[45] That amounts to 2.4% of the total cost of health care costs.

Though the authors of this study regard 2.4% as a relatively small fraction of total health care costs, they nevertheless consider 55.6 billion dollars no trivial amount of money. Therefore, they offer a proposal for tort reform to reduce the costs of the medical liability system. They opt for federally imposed collateral-source offsets, combined with moving toward universal health care coverage. Collateral-source offsets mean that costs covered by their health insurance cannot be recovered by malpractice plaintiffs. That is intended to lower liability costs. And increasing the number of people who have health insurance reduces the number of people who would otherwise be seeking to be reimbursed for what they had to pay for treating the injuries that they suffered. Furthermore, the authors contend that granting insurance coverage to those presently uninsured may reduce their need to file malpractice claims for medical expenses arising from injuries caused by malpractice since these would have been paid by their insurance companies.

There are states that have instituted various kinds of tort reform. These include collateral-course offsets and capping total liability. Taking note of the ineffective results of these and other reforms, a study pinpointed the reason why this has proved to be the case: these reforms have not been able to reduce the fears physicians have for malpractice lawsuits.[46] The study found that what physicians dread is being

---

[44] Ibid, Assuming a 30 percent reduction in malpractice insurance premiums would yield an 0.4 percent decrease in total health care costs.

[45] Michelle M. Mello, Amitabh Chandra, Atul A. Gawande, and David M. Studdert, "National Costs of the Medical Liability System," Health Affairs 29:9, September 2010, 1585–1592.

[46] Emily R. Carrier James D. Reschovsky Michelle M. Mello, Ralph C. Mayrell, and David Katz, "Physicians' Fears of Malpractice Are Not Assuaged by Tort Reforms," Health Affairs, 29:9, September 2010, 1585–1592.

dragged into courts and subjected to what they perceive to be unfair, arbitrary, and adversarial proceedings. Also, the result of these court cases is that the physician's reputation is sullied regardless of the outcome. The authors conclude that comprehensive liability reform is essential to reduce defensive medicine.[47]

What is more, as the physician Norton Hadler has noted, it is not only physicians but also patients who suffer in the American tort system because of what drives it:

> Many stakeholders stand to lose if the victim regains full health: the attorneys on both sides lose income, the insurance companies cannot justify escalating premiums, and the healthcare providers of every ilk involved in clinical recourse, valuing loss, and the determination of residual disability lose their raison d'etre sooner. The poor victim is under subliminal if not overt pressure to enhance his or her illness perceptions.[48]

This kind of pressure is especially egregious since an injury due to medical intervention involves pain and suffering that physicians normally seek to alleviate in a timely fashion and patients normally seek to have alleviated rather than prolonged. But prolonging the patient's malaise is precisely what happens in courtroom deliberations, as lawyers spar and then juries deliberate to come up with a verdict. There are, then, substantial reasons for avoiding a lawsuit for assessing whether a bad outcome for a patient involves malpractice on the part of a physician.

A process for doing just that has already been instituted by a number of large institutions, in particular hospitals and clinics.[49] These institutions are establishing independent panels comprised of peers, selected largely from the community. These panels are charged with the task of deciding causation, and if they find malpractice to be the cause, deciding fair compensation for the patient. The cases of untoward outcomes they are assessing have been identified by staff or patients as instances of possible malpractice.

The standard these panels use to decide whether what caused an untoward outcome is an instance of malpractice is based on the relevant and valid scientific evidence that either justifies or does not justify a physician's actions in the case being reviewed. The standard is akin to the "Daubert Standard of Expert Opinion," formulated by the U.S. Supreme Court in Daubert v. Merrell Dow Pharmaceuticals, Inc. (509U.S. 579, 584–87), a rule that scientific evidence presented to courts, judges, or juries be relevant and reliable: appeals by experts that what they present is "generally accepted" will no longer suffice.[50]

The work of these panels has had salutary results. Where the process described above has been adopted, institutions have witnessed a decrease in the cost of malpractice and something to celebrate even more, the occurrences of malpractice have decreased as well. It would seem that tort reform for dealing with medical

---

[47] Ibid, 1591.
[48] Hadler, Citizen Patient, 116.
[49] Ibid, 117–118. Type I Medical Malpractice for Hadler refers to cases of untoward results from necessary medical interventions: Type II refers to unnecessary interventions that Hadler would like to include in considerations of malpractice.
[50] See ibid., 92–93 for Hadler's explanation and discussion of the Supreme Court's decision and the "Daubert Rule."

malpractice could profitably consist of legalizing this process a number of medical institutions have already put in place.

However, something more would be required to assure the validity and fairness of such a policy. There may be instances in which an injured party feels strongly that the compensation being offered them is insufficient. Surely there should be an independent, independently selected lawyer available to mediate such a conflict and have that mediation be binding on both parties to the conflict.

As Mary Ann Glendon, while a law professor at Harvard Law School argued in her masterful study of American legal practices, lawyers should, first and foremost, be peacemakers and harm evasive, settling disputes by mediation, and finding ways to keep people out of court when it is legally appropriate and morally called for. She found the current resort to litigation increasingly excessive and too often harmful.[51]

7.2 What Physicians Are Doing and Advising Physicians to Do to Achieve the High Moral Ground in the Provision of Medical Care

Physicians make decisions that add up to 80% of all healthcare expenditures.[52] Clearly, what physicians do matters. In my efforts to provide evidence for the central claim of this book that a just health care system is sustainable, I have documented considerable examples of medical decisions and practices that are costly, unnecessary, harmful, and scientifically and ethically unjustifiable. I did this without any intention to portray U.S. physicians generally in a bad light. Unfortunately, many physicians are unaware of the extent to which their medical decisions are guided by FDA approved, but scientifically and ethically unsound studies, financed by pharmaceutical and medical device manufacturers. This lack of awareness is understandable since a great many of such studies appear in leading medical journals. Some bear the name of a leading medical expert who lends credence to the claims being made on behalf of the medical benefits of the product that was studied though the research was not carried out, or even written up, by the expert who is named as the author of the article in question.

Many physicians are also relatively unaware of how much they are being influenced by the presence of personnel from drug companies in their medical school, their offices, and in their continuing education. As the reader knows, all of these matters have been thoroughly documented and illustrated in Chaps. 4, 5 and 6.

---

[51] Mary Ann Glendon, A Nation Under Lawyers: the Impoverishment of Political Discourse (New York: the Free Press, 1991).
[52] Abramson, Overdosed America, 111.

7.2.1 Developing Advisories Designed to Curb Scientifically Unjustifiable Harmful and Costly Interventions

Fortunately, the "Choosing Wisely Movement," described in Chap. 4, reflects a rapidly growing awareness that scientifically unsound harmful and costly medical interventions are far too common and should not be tolerated. This movement is breathing new life into practicing the traditional professional medical ethic.

As the reader will recall, the "Choosing Wisely Movement" consists of an expanding number of professional medical organizations representing various areas of medical specialization. These organizations are issuing recommendations to make scientifically sound medical decisions that have the effect of avoiding interventions specified as harmful, costly, and non-beneficial. The organizations involved and the recommendations being promulgated are increasing at quite a pace. Rather than offer a list of these that will likely be incomplete by the time this book is published, I invite the reader to visit www.choosingwisely.org. Obtaining these lists will provide vital guidance to both physicians and patients as they decide which medical interventions to favor and which to decline.

As significant and helpful as these advisories of the "Choosing Wisely Movement" are, they do not, at least not yet, address some of the additional changes in physician behavior and thinking needed to replace still more practices that cannot be justified scientifically, ethically, and financially. We turn now to consider some of these changes and how they are to be realized.

7.2.2 Putting Interventions to Prevent Diseases on a Scientifically and Ethically Sound Footing

There is a widespread belief among physicians that it is beneficial to detect health problems as early as possible. Welch refers to this belief as a "paradigm" adopted by medicine.[53] An early diagnosis of someone experiencing symptoms is certainly advantageous, particularly if they are symptoms of an illness not expected to go away without intervening medically. But early diagnosis of people who are well, can and often does, lead to overdiagnosis and unnecessary treatments.

However, as Welch points out, it is difficult to raise critical questions about early diagnosis because it has come to be regarded as synonymous with preventive medicine. Thus, since preventive medicine is widely held to be unambiguously good, early detection has the same reputation as an unambiguous good.[54] Some would contend that early detection seeks to find disease, not prevent it. Welch has a more nuanced view. Early detection is looking for abnormalities early on to prevent what may result from them in the future. Since many individuals harbor abnormalities

---

[53] Welch, Overdiagnosed, 182,
[54] Ibid, 190.

that will never bring about any deleterious consequences, being subjected to this kind of preventive care is, Welch notes, "the fastest way to become sick."[55]

How does that happen? Consider a case Welch experienced with one of his patients. Mr. Roberts was diagnosed by Welch to be mildly diabetic on the basis of a slightly elevated level of blood sugars.[56] That prompted Welch to treat Mr. Roberts with glyburide—a drug that lowers blood sugar. It worked too well. Six weeks later, Mr. Roberts blacked out while driving from too little blood sugar. He suffered a broken neck from which he fortunately recovered. Welch took him off the drug and Mr. Roberts received no more treatment for diabetes. At the time Welch wrote about this example of overdiagnosis followed by an unnecessary treatment, Mr. Roberts had not suffered any symptoms of diabetes for 16 years: he was 74 at the time he was treated and treatment was stopped; he was now 90!

Ironically one very important way to prevent illnesses is to stop treating people who do not have any symptoms of the illness in question. At the same time, individuals are being judged as abnormal by standards that make it very difficult, for example, to achieve low enough levels of blood pressure, cholesterol, and blood sugar and high enough bone density to be regarded as normal. Hence, the result of moving increasingly in this direction has led to labeling additional millions of people as at risk of experiencing heart attacks and strokes, as having diabetes 2, and as suffering from osteoporosis and needing treatment to prevent falls. (See footnote 64 above for the numbers involved.)

Alas, individuals labeled as being at risk for these illnesses are most commonly treated with drugs: for all of these instances, there are superior alternative interventions, more efficacious, much less costly and free of the harmful side effects of drugs. These matters are documented in Chap. 4.

These alternatives appear in Abramson's list for living a healthy way of life. This list includes his version of the Mediterranean diet, exercise, maintaining normal weight, avoiding tobacco, consuming alcohol moderately or abstaining entirely.[57] It should be noted, that in his discussion of osteoporosis in Chap. 4, Abramson has some additional specific dietary recommendations to prevent it, namely vitamin D, calcium, and magnesium to strengthen bones.

We see, then, that the advice physicians like Welch and Abramson are giving other physicians to prevent the development of illnesses and their adverse symptoms, focus on promoting and facilitating practices that contribute to the health and wellbeing of their patients. Physicians are to advocate especially a healthy diet and exercise, and the avoidance of toxic and addictive substances like tobacco. Physicians who help patients develop a healthy lifestyle are also helping them avoid

---

[55] Ibid. On page 23 the millions and increasing millions who are made sick is documented in Table 2.1.

[56] Ibid, 16–17.

[57] Abramson, Overdosed America, 237–239. See also R. Estruch, E. Ross, J. Solas Savado, et alia, "Primary prevention of cardiovascular disease with Mediterranean diet," New England Journal of Medicine, 2013:308 (14), 1279–1290.

the risks of being harmed by drugs and other risky medical interventions. Doing so is in line with a key tenet of the traditional medical ethic: first of all, do no harm.

7.2.3 Preventing Rampant Diagnostic Inflation and Misguided Treatment of Mental Disorders

I will not repeat my account of the very extensive diagnostic inflation, and misguided treatments of mental illnesses, and their very deleterious consequences: that account is documented and illustrated in Chaps. 5 and 6.[58] What now follows is advice from a noted psychiatrist that indicates ways to curb such practices for the benefit of patients and the very practice of psychiatry.

7.2.3.1 Stepped Diagnosis to Curb Diagnostic Inflation

The psychiatrist Frances regards the way psychiatric diagnosis is currently carried out as unacceptable. Mental health clinicians and primary care physicians need to be taught not to make a diagnosis on the first visit, based too often on a brief interview and an inadequate amount of information. What is learned in a second visit is likely to differ from that first visit. Frances suggests a total of six visits to get things straight so as either to make a diagnosis or allow time for what it takes for some people to get better naturally. What follows now is his "stepped diagnosis:"

STEP 1 – Gather baseline data
STEP 2 – Normalize problems: take them seriously, but reformulate positively as expectable responses to the inevitable stress in life.
STEP 3 – Watchful waiting: continued assessment with no pretense of a definitive diagnosis or active treatment.
STEP 4 – Minimal interventions: education, books, computer-aided self-help.
STEP 5 – Brief counseling.
STEP 6 – Definitive diagnosis and treatment.[59]

This process of arriving at a diagnosis is also a recipe for altogether avoiding labeling someone as mentally ill. That is very important because the label, especially when unwarranted, exposes individuals to some extremely harmful side effects. After all, as we learned in Chap. 5, antidepressants come with a risk of suicide, and violent ideation and behavior and more, and antipsychotic drugs can lead, and have led, to deaths, particularly of very young children and the elderly. Furthermore, being labeled as mentally ill also undermines morally responsible behavior: it risks being an excuse for inexcusable behavior.

---

[58] For that documentation, see Frances, Saving Normal, 103–106 and 77–113.
[59] Ibid, 222.

Stepped diagnosis provides the opportunity and the time to present a diagnosis of a mental disorder; time itself allows our natural capacity to heal ourselves to function. Hippocratics, with their dedication to avoid harming their patients, took advantage of our natural capacity to heal ourselves.

There are additional alternatives to diagnoses of mental disorders that curb overdiagnosis and overtreatment. There also are alternatives to treatments using drugs.

7.2.3.2 Alternatives to Unnecessary Diagnoses and Drug Treatments of Mental Illnesses

For a considerable number of people with mild and transient symptoms, antidepressants (SSRIs) are nothing more than very high-priced placebos with harmful side effects. There are alternatives to prescribing psychotropic drugs and to being considered mentally ill: these include "natural resilience, exercise, family and social support, and psychotherapy."[60]

Psychotherapy is as efficacious as drugs in dealing with mild to moderately severe problems.[61] Frances laments the declining practice of psychotherapy in the U.S. He applauds Japan's national effort to switch from medication based therapy to cognitive therapy since it is proving cost-effective.[62]

But initiating stepped diagnosis opens up the opportunity to prevent someone from being stuck with the label of being mentally ill and thus avoid psychotherapy as well as drugs and their side effects. Pointing out the normality of what an individual is experiencing is what that individual is offered in Step 2: that kind of reassurance may happily trigger the placebo effect.

Early on, in carrying out stepped diagnosis with a patient, there is also the opening to suggest what exercise and diet can do to promote well being and relieve distress. Indeed, Abramson comes to the conclusion that depression, if not very severe, should be regarded as exercise deficiency.[63] Furthermore, studies show that brief, 15 min sessions with a family physician may dispel the symptoms of depression and of the so-called social anxiety disorder (SAD).[64] That could also occur in Step 5 or earlier in the use of stepped diagnosis, so that no diagnosis and no treatment of depression, SAD, or any other mental disorder need occur.

---

[60] Ibid, 157.
[61] Ibid, 108.
[62] Ibid.
[63] Abramson, Overdosed America, 232–233.
[64] Ibid.

7.2.3.3 Avoid Diagnosing and Treating Normal Feelings and Behavior as Mental Illnesses

Regrettably, as documented in Chaps. 5 and 6, falsely and excessively diagnosing and treating people or mentally ill is not confined to stretching and distorting existing categories and criteria of what counts as a mental illness. The authors of the newest versions of psychiatric guidelines (DSM-5) have created new categories of mental disorders for feelings and behaviors that, as Frances contends, should as such be viewed as normal, and should not, therefore, be diagnosed and treated as mental disorders. These are normal feelings and behaviors turned into mental disorders by DSM-5: temper tantrums; forgetfulness of normal aging; gluttony; mourning and grief; and behavioral addictions, that is, normal habits and activities that are pleasurably and repetitiously enjoyed.

None of these alleged mental disorders are defined precisely enough to preclude mislabeling many individuals that are currently deemed to be normal. No effective treatments for these purported mental disorders exist. To the extent to which those categories of mental illness are used for diagnosing and treating individuals, the overall consequences will be "overdiagnosis, unnecessary stigmas, overtreatment, a misallocation of resources, and a negative impact in the way we see ourselves as individuals and as a society."[65] Frances is saying to primary care physicians and psychiatrists: don't use, just ignore these newly created categories of mental illnesses found in DSM-5. These categories are not needed to deal with instances of gross departures from normal feelings and behavior.

7.2.4 *Dealing Ethically with Conflicts of Interest*

Hadler considers the U.S. health care system to be ethically bankrupt: what makes it so is conflict of interest.[66] As rich sources of conflict of interest, he points to certain kinds of arrangements between individual practitioners and representatives of drug and device manufacturers marketing their products. These drug and device representatives visit physicians at the site of their practices, laden with free samples of their products, descriptions of their merits and use, and dispensing gifts and free meals. This kind of direct marketing has been banned from some institutions but many private practices remain open to them.

Hadler is emphatic about advising physicians to shun these sorts of conflict of interest. To be shunned as well, is drug company sponsored and financed continuing education. As his guide to dealing ethically with conflicts of interest, Hadler has formulated this moral rule: "No physician should knowingly enter into any arrangement that might compromise trustworthiness in any treatment act."[67] The implication

---

[65] Frances, Saving Normal, 74. For a full discussion of these issues, see pages 170–192.
[66] Hadler, Citizen Patient, 5.
[67] Ibid, 8.

of this rule is that: "any physician who does not share this moral judgment is a compromised healer."[68] Furthermore, continuing any such conflicts of interest cannot be justified through disclosing them: rather, Hadler contends, the very need to disclose them "signifies immorality."[69]

Frances, like Hadler, thinks that physicians should not have their clinical decisions influenced by the present kinds of marketing activities carried out by drug companies that Hadler described. To accomplish that, he would have the federal government ban these marketing tactics.[70]

There are some further conflicts of interest that persist even in the face of opposition by leading physicians. In 2001, the editors of major medical journals issued a joint statement recommending that "researchers have control over their data, analysis and publication of their work."[71] Why would it be necessary to formulate and publicly issue a statement asking for behavior that everyone should be able to expect from anyone engaged in scientific research? The reason is that the heads of departments in many academic medical centers have been willing to sign off on contracts for studies being conducted and financed by drug companies since they bring in revenue for the department and recognition for a department member, even when these contracts cede control over the data, analysis, and publication of these studies to the drug company. A study conducted a year later found that it was rare to find academic institutions who were following the recommendations of these medical journal editors who appropriately wished to be certain that no study they reviewed for publication lacked scientific integrity and was flawed in ways they could not be able to discover.

Drug companies pursue their own best interests still another way. They hire ghost writers to write up the original draft of a completed trial. This draft is submitted to the drug company and then from there it is given to the author of record, some notably busy physician delighted to be spared the labor of coming up with a first draft. By these means, the drug company has, from the outset, infused its perspective into how the results are to be seen. Drug companies seek and retain only those doctors they can influence to have their name on the publication. The article in the New York Times describing all this calls the result "marketing masquerading as science."[72] A study found that ghostwritten articles in peer-reviewed medical journals constitute 11% of their published articles; and 19% of all articles listed as authors individuals who contributed too little to the research to qualify as authors. Obviously, Abramson cites these practices as ones that physicians should shun. To engage in these practices is to put one's own interests above the interests of patients, science, and tax payers. I would add, to engage in such practices is a gross failure to meet the standards both of medicine as a profession and of ethics at its very core.

---

[68] Ibid.
[69] Ibid, 7.
[70] Frances, Saving Normal, 212–213.
[71] Abramson, Overdosed America, 105–106.
[72] Ibid, 106–107.

At this point one might well ask whether having ties to manufacturers of drugs and medical products, particularly by receiving money from them, necessarily compromises physicians as those who are always to act in the best interests of their patients. That need not be the case. Abramson presents an example.[73]

William Applegate, a physician, consented to do research sponsored and financed by a drug company. The data he and his colleagues had gathered showed that the new drug they had studied had more side effects than the drug in use for the same purpose. Before he and the company sat down to review the data, Applegate was offered 30 thousand dollars a year to be a consultant. That prompted Applegate to suspect that the company was trying to buy from him a favorable opinion of the drug he had studied: he turned down the offer. So then, when the company twice offered grants to Applegate's research center, each time inquiring whether he had changed his mind about the conclusions of the study, Applegate refused both overtures and he and three of his colleagues resigned from the study. They resigned in order to reject outright the drug company's unethical behavior and intentions and thereby protect the best interests of the patients by shielding them from using a drug they did not need and from a drug more harmful than the drug they could use instead, one already in use. In the end, because of these resignations, a fair result was presented one that Applegate and his colleagues could and did endorse.

This example illustrates two things at once: First, it is possible to accept money from drug companies while rejecting any contractual arrangements with, or any input from these companies that would undermine the scientific character and honest presentations of the results of the research; secondly, there are physicians who do and can accept money from drug companies who do and can refrain from unethical conduct or moral compromises.

In Chap. 4, I made the point that those who have financial ties to drug companies, have a moral obligation that some physicians are not honoring. I refer to the willingness to serve on advisory committees for the FDA making decisions about drugs manufactured by the companies with which they have monetary ties. I refer also to those who serve on a panel of experts to set guidelines for the use of drugs issued by companies with which they have financial ties. A judge has a moral obligation to recuse himself or herself from any case in which a particular verdict would be beneficial or deleterious to his or her interests. These judges of what drugs to use and when, have the very same moral obligation. A reformed FDA, as I argued, would compel them to meet that moral obligation.

Conflicts of interest are hard to avoid and not easily dealt with. I side with Hadler in contending that simply disclosing them is never a sufficient way to handle them. Some of them, as the physicians I have cited advise, should be shunned altogether. And those that are not necessarily to be shunned, obligate physicians to behave like Applegate, acting always in ways that protect the best interests of patients. They also should not accept any money for any activities or for holding any opinions or making any judgments that are scientifically unjustified.

---

[73] Ibid, 108–109. There are physicians who do not accept funding from drug and medical device companies at all. See Marty Makary Unaccountable, 139.

7.2.5 Correcting Medical Errors

Medical errors are a serious problem and physicians are in a position to reduce them. Twenty-five percent of hospitalized patients experience medical errors.[74] The results of preventable errors by physicians not only involve a great deal of suffering, but according to one study, result in hundreds of thousand of deaths as well every year.[75]

These medical errors are costly. A very costly one occurred when the University of Chicago Hospital failed to detect that the organs extracted from the donor was Hepatitis C-positive and HIV-positive. They gave these organs to four different patients who were free of these diseases. Now these four patients have to live with HIV and Hepatitis C and take medications expensive enough to make this medical error a multimillion dollar mistake.[76]

This disturbing information is reported in the book published by Marty Makary who is a leading surgeon at Johns Hopkins Hospital and an associate professor of health policy at Johns Hopkins School of Public Health.[77] He is very concerned about the tragic consequences of medical errors and has a research focus studying the prevalence of medical errors and ways to prevent them. What follows now is helpful advice to physicians to help them and their hospitals reduce medical mistakes, advice solidly based on his research and his own experience.

7.2.5.1 Practice and Facilitate Teamwork

For reducing medical errors, Makary's research demonstrates that:

> When teamwork is good, the hospitals have good outcomes. When it is poor, hospitals have worse outcomes.[78]

Makary describes how this works in his own practice. As he met with a patient scheduled to have fluid removed from one of his lungs, he was called to attend to a complication threatening a patient in the ICU. That disruption meant that the preparation of his patient for surgery was undertaken and finished by a new intern assigned to work with and learn from him. The preparation was well done, on the right side of the patient, with a window of skin properly sterilized and laid bare, surrounded by sterile towels.

---

[74] C.P. Landrigan, "Temporal Trends in Rates of Patient Harm Resulting from Medical Care," New England Journal of Medicine, 363, no. 22, 2010, 2124–34.

[75] Committee on Quality Health Care in American and Institute of Medicine with eds, L.T. Kohn, J.M. Corrigan, M.S. Donaldson, To Err is Human: Building a Safer Health System (Washington, D.C: National Academy Press, 2000).

[76] Denise Grady, "Four Transplant Recipients Contract HIV," New York Times, November 13, 2007.

[77] Marty Makary, Unaccountable.

[78] Ibid, 20.

Just as Makary was about to make his incision the nurse spoke up, asking whether the surgery was to be on the right or left side of the patient, noting that the left side of the chest was what was called for. Makary was clearly taken aback, his heart racing. He thanked the nurse for speaking up and preventing what he called a "close call." Because the nurse had done her homework and had the courage to speak up, the surgery was done on the correct side and a medical error was avoided.

Makary points out that this was not the first time he was saved from a catastrophe by a nurse. Creating a culture in which all fellow workers are encouraged to express any of their safety concerns improves safety; failure of staff members to speak up compromises safety.[79]

Teaming up with a colleague, social psychologist Bryan Sexton, Makary devised a survey that tested hospitals for their safety culture, in particular the extent to which health care workers felt comfortable communicating their safety concerns, and whether they rated teamwork in their setting as good. They undertook their "Hopkins Safety Culture Study," enlisting sixty reputable U.S. hospitals to administer their survey to all their employees. The results showed considerable variation in the safety culture of these hospitals, as well as among their departments: for example, surgery could have a perfect teamwork culture, while ob-gyn rated poorly for teamwork with a given hospital. There were hospitals in which fewer than 20% reported good teamwork, and at the same time, there were hospitals in which 99% of the staff considered teamwork of their hospital good. What is important to note is that the ratings of teamwork correlate with infection rates and patient outcomes in these hospitals.[80]

Makary and his associate, using survey data from this and subsequent studies, also discovered that a highly rated safety culture corresponds with a low surgical complication rate and conversely, a poor rating for safety corresponds with a poor rating for surgical complications. Safety attitudes and surgical complication rates by hospitals are similarly distributed.[81]

What Makary has found out is that his own clinical experiences, that good teamwork and a good safety culture prevent errors and improve the quality of care, are objectively verifiable, generalizable phenomena present in all hospital settings. That enables Makary confidently to draw on his own practices to advise physicians as to how they can create teamwork and a culture of safety.

Makary urges that physicians cultivate good relations with nurses and the whole staff. Nurses have saved him from many a medical error when he first began to practice medicine and ever since. Physicians should know the names of the nurses: nurses report that it is a sign of respect. It is important as well to be pleasant and not to lose one's temper.

Physicians should encourage the whole staff and their fellow physicians to be on the lookout for possible mistakes and to give voice to their every concern about safety and how to assure it. In turn, physicians should express their thankfulness for

---

[79] Ibid, 170–172.
[80] Ibid, 23–27.
[81] Ibid, 90–92.

being corrected to anyone who corrects them before they make a mistake, and they should not react with defensiveness or anger.

Physicians should not fail to refer their patients to anyone among their colleagues, or if necessary, to physicians outside their hospitals if they know their patients will be better and more safely served by such a physician. Makary stresses the need to do this since it is not practiced nearly enough and patients suffer because of that. He devotes a whole chapter of his book documenting how vital a practice it is.[82]

7.2.5.2 Practice and Facilitate Involving Patients in Ways that Help Avoid Medical Errors

One significant source of errors are the mistakes that can occur in up to half of all medical records. Makary has found that these mistakes can be considerably reduced by sharing with patients what is being put into their records. He does this by having them listen as he dictates his notes, record them into the medical record. He tells them what he is about to do and asks them to correct anything he has missed or that doesn't sound right; they gladly consent. They do sometimes ask for changes, sometimes they are correcting what they said, and sometimes they correct what Makary wrote in his notes. He always thanks them and dictates "correction" into the record. There are other hospitals besides Johns Hopkins now trying out what Makary calls "open notes."[83]

7.3 A Concluding Epilogue

For centuries since the time of Hippocrates and the Hippocratics in the fifth century BCE, the ethical outlook I have dubbed "the traditional medical ethic" has guided the decisions, practices, and thinking of physicians. That outlook reflects what I have portrayed as demands of justice in its generic sense, namely what we owe one another as human beings. As these demands pertain to medical care, what these moral demands add up to is that: first, it is unjust to deny anyone their genuine need for medical care regardless of whether they lack the means to pay for it sufficiently or at all; secondly, the demands of justice are such that providing everyone with the medical care they genuinely need is financially sustainable.

The urge to provide medical care to those who genuinely need it is fueled by natural, emotional impulses and abilities neuroscientists have discovered to be emanating from our brains. As human beings, we have inhibitions against harming ourselves and others, and the ability to detect when someone is suffering (empathic accuracy), sympathize with that individual, and be empathically driven by

---
[82] Ibid, 57–74.
[83] Ibid, 181–184.

compassion to prevent, relieve, or rescue someone we perceive to be suffering or in danger of suffering, or of losing their very lives. Indeed, people who are cited for heroic, life-saving rescues are often quoted in newspapers as saying they are not heroes, rather they simply did what anyone would do.

Our inhibitions against harming one another and our proclivities to be compassionate are specifically expressed in what I call the moral requisites of individual and communal life. The inhibitions against harming one another include refraining from killing, lying and stealing, that they are deemed to be moral requisites of individual and communal life is evident in laws against murder, deceitfulness, such as breaking contracts and fraud, and theft. Voluntarily refraining from these harms is expected; certain deviations from these moral demands are too dangerous to permit.

These moral demands on us are not only inhibitions but also proclivities driven by our empathic emotions. We are led not only to avoid killing but also to protect and save lives; not only to avoid lying but also to tell the truth; not only to avoid stealing but to protect one another's property.

Attaining reliable knowledge depends upon truth-telling. Scientific discovery of truth is only possible if we shun deceitfulness of every kind, and if we communicate our findings truthfully in every regard. Informed consent to medical interventions cannot occur if truthfulness is at all compromised.

Another proclivity of our empathic emotions is nurture, and nurture is legally enforced. For example, parents are subject to punishment for avoidable failures in nurture that prove harmful to their children. Similarly, avoidable failures to nurture patients in ways that harm them can trigger legal sanctions or lawsuits.

A behavior, practice, or policy is ethically and scientifically unjustifiable when they violate one or more of these moral demands of justice. Examples of these I discuss in Chap. 6. Then in this concluding chapter, I have offered some suggestions to show that there are ways to reduce what I have identified as unethical within the U.S. health care system. In whatever way these unethical behaviors, practices and policies are diminished as much as possible, it is urgent since failing to do so perpetuates the current unsustainability of the U.S. health care system.

What, however, are the prospects for achieving a sustainable and a just U.S. health care system? I have illustrated the efforts and proposals of the physicians I have cited as well as those of the Choosing Wisely Movement. These initiatives and insights if widely acted upon and incorporated into federal health policy would save billions of dollars and also eliminate a great many harmful unethical and unscientific practices. It is too early to predict how successful and how far these efforts will go.

However, there is one essential moral demand that is not being addressed. It is the hallmark of the traditional medical ethic. It is the moral demand, emanating from our empathic emotions, that of sharing the revenue raised from providing medical services to pay for the medical services of those who cannot pay sufficiently or at all for them.

Health care is now so expensive that the fees of physicians cannot now begin to provide for all their patients in need of financial assistance as once they could and did for centuries. However, what is exceedingly expensive for physicians, patients and hospitals is what generates the very money that would allow for free medical care for those who require it. For those monetary resources to be spent on patient care, it would be necessary that the manufacturers of drugs, medical devices and medical technology, the big wealthy hospitals, health insurers and wealthy physicians, like the CEOs of highly profitable hospitals, to adopt and practice the traditional professional medical ethic: that means sharing their enormous profits and compensation to assure that no one lacking the resources for receiving medical care is denied it.

In essence, this way of generating revenue to make available free and practically free medical care so that all genuine medical needs for it are met for everyone is a matter of adapting the practice of the traditional medical ethic to the current situation in which medicine is taking place. When such a health care policy is combined with policies designed to curb the kinds of ethically and scientifically unjustifiable behavior in the present U.S. health care system, you achieve justice, namely equally and fully provided quality medical care for everyone and you do so sustainably.

As I write this, the U.S. Senate has health care policy on its agenda. Although senators are concerned about assisting people financially who need such assistance to obtain medical care, and to accomplish this sustainably, they are not proposing the two pathways I suggested above are essential if sustainability is to be attained. Instead, sustainability is approached as a matter of deciding what of the federal government's revenue can affordably be spent on the provision of health care. The federal government is not on a path to a just and sustainable health care system: the traditional professional medical ethic is not guiding its health policy deliberations.

The traditionally guiding ethic of the medical profession is also not the focus of the phenomenally thriving commercial enterprises that manufacture and market a whole array of medicines, devices and technology very much in use in dispensing medical care: the focus is on making money. The empathy that informs the practice of the traditional medical ethic is for too many in these businesses utterly overwhelmed by what can only be described as an insatiable thirst for profit. How else can one explain the unethical conduct that, at times, is nothing short of merciless. These exceedingly profitable manufacturers of medical products have the money to wield enormous influence in every sector of the U.S. health care system. That influence, unfortunately, includes heavy lobbying, especially by drug companies, to keep Congress satisfied with the status quo.

When the physician Norton Hadler observes this pervasive influence on the various participants in the U.S. health care system he is led to declare that system to be "essentially ethically bankrupt" and "that essence" to be "conflict of interest."[84] For him that means that:

> The American health-care system is as entrenched and wealthy as it is perverse. It is also clearly unsustainable. It is hell-bent for collapse, and its leadership is much more committed to and adept at self-service than it is farsighted or ethical. The result will be turbulent, and

---

[84] Halder, Citizen Patient, 5.

many a lovely servant to the ill and many an ill patient will be sorely served during this time.[85]

In Hadler's judgment, it is only after the U.S. health care system has become so painfully and thoroughly unsatisfactory for patients and physicians alike that the great bulk of the citizenry will demand that it be remedied.

In my view, however, any changes in the U.S. health care system will not achieve justice or become sustainable unless the necessary ethical orientation is awakened. That begins by a return to the basics. As they dispense t heir services, physicians are supposed to always act in the best interest of patients. That fundamental guide to conduct ought to guide every person and every entity that provides for the care of patients in any way, including what products they offer or make for use in dispensing medical care. For the Hippocratic ethic, the profit made from serving patients is the means utilized to treat everyone, regardless of their financial situation. Profit made by companies making products for the care of patients should serve as a means to assure that no poverty-stricken patient is denied medical are.

Shockingly, the greed that undermines the morality of commercial enterprises is similarly corrupting some medical caregivers as well. I learned today, that among 412 people charged with defrauding Medicare, are almost 115 doctors, nurses, and other medical professionals, such as pharmacists. Of those charged, more than 120 people are cited for illegally prescribing and distributing opioids and other dangerous narcotics: this they have been doing while approximately 91 Americans are dying daily from opioid-related overdosing. Over and above the hundreds charged for defrauding Medicare, are 300 medical caregivers, including doctors, nurses, and pharmacists, for whom suspensions are being sought by the Department of Health and Human Services.[86]

Once again, we see that when the love of money suppresses our empathic emotions, the resulting behavior can reach a point of sheer mercilessness. In those instances, fraud and violations of professional conduct are regarded as failures in moral responsibilities that should be legally enforced. Doing this treats these moral responsibilities as moral requisites of individual and communal life. That is the orientation, the understanding of morality needed to break loose from and overcome the moral bankruptcy of the U.S. health care system. Serving the best interests of patients, rescuing individuals from suffering, illness, and death, is what justice demands. And these moral demands, like the demand not to commit fraud, are moral requisites of individual and communal life that ought to be legally enforced as deemed necessary. The just provision of medical care can only occur in a system that denies nobody the medical care that they genuinely need, and that shares the revenue medical care generates. Only then will justice be realized and sustained.

---

[85] Ibid, 222.
[86] Reuters, "Doctors, Nurses Among Hundreds Charged With Defrauding Health Programs," New Hampshire Union Leader, July 14, 2017, page A8.

7.3 A Concluding Epilogue

What in this book I am asking of participants in the U.S. health care system is nothing more than to let nothing deter them from acting in accord with the kind of behavior that flows from the empathic emotions we naturally share as human beings. Menno Simons understood that insatiable thirst for profit and compensation thwarts justice. Robert Frost understood that "nothing can make the unjust just but mercy." I agree with them. And, at book-length, I have given my reasons why.

Bibliography

Abramson, John. 2004. *Overdosed America: The Broken Promise of American Medicine*. New York: Harper Collins.
Allport, Gordon. 1955. *Becoming*. New Haven: Yale University Press.
Asch, Solomon. 1952. *Social Psychology*. Englewood Cliff: Prentice Hall.
Bendapudi, N.M., L.L. Berry, K. Frey, J.T. Parish, and W. Rayburn. 2006, April. Patients' Perspectives on Ideal Physician Behavior. *Mayo Clinic Proceedings* 81 (3): 338–344.
Berkrot, B., and R. Pierson. 2012, February. FDA Adds Diabetes, Memory Loss to Statins. *Union Leader* 29: 87.
Brill, Stephen. 2013, March. Bitter Bill: How Outrageous, Pricing and Egregious Profits are Destroying Our Healthcare. *Time* 4: 16–55.
Callahan, Daniel. 2008, March. Curbing Medical Costs. *America* 10: 9–12.
Carrier, Emily R., James D. Reschovsky, Michelle M. Mello, Ralph C. Mayrell, and David Katz. 2010, September. Physicians' Fear of Malpractice are not Assuaged by Tort Reforms. *Health Affair*: 1585–1592.
Cassel, Christine K., and James A. Guest. 2012, May 2. Choosing Wisely: Helping: Physicians and Patients Make Smart Decisions About their Core. *JAMA* 307: 1801–1802.
Chernew, Michael E., Lindsay Sabik, Amitabh Chandra, and Joseph P. Newhouse. 2010, January. Ensuring the Fiscal Sustainability of Health Care Reform. *New England Journal of Medicine* 7: 1–3.
Colwill, J.M., J.M. Cultice, and R.L. Kruse. 2008, May–June. Will Generalist Physician Supply Meet Demands of an Increasing and Aging Population? *Health Affair, Millwood* 27 (3): 232–241.
Conrad, Peter. 2007. *The Medicalization of Society: On the Transform of Human Conditions into Treatable Disorders*. Baltimore: The Johns Hopkins University Press.
Damasio, Antonio. 2005. *Descartes' Error: Emotion, Reason and the Human Brain*. New York: Penguin Books.
Daniels, Norman. 2008. *Just Health: Meeting Health Needs Fairly*. Cambridge, MA: Harvard University Press.
Dyck, Arthur J. 2005). *Rethinking Rights and Responsibilities: The Moral Bonds of Community*. Revised ed. Washington, DC: Georgetown University Press.
———. 1977. *On Human Care*. Nashville: Abingdon.
Firth, Roderick. 1952, March. Ethical Absolution and the Ideal Observer. *Philosophy and Phenomenological Research*: 317–345.

Frances, Allen. 2013. *Saving Normal: An Insider's Revolt Against Out of Control Psychiatric Diagnosis, DSM-5, Big Pharma and the Medicalization of Ordinary Life*. New York: Harper Collins.
Frankena, William K. 1973. *Ethics*. Englewood Cliff: Prentice Hall.
Frost, Robert. 1979. A Masque of Mercy. In *The Poetry of Robert Frost: The Collected Poems, Complete and Unabridged*, ed. Edward Connery Lathem, 493–521. New York: Holt, Rinehart, and Winston.
Glendon, Mary Ann. 1993. *A Nation Under Lawyer: The Impoverishment of Political Discourse*. New York: The Free Press.
———. 1991. *Rights Talk*. New York: The Free Press.
Goldacre, Ben. 2012. *Bad Pharma: How Drug Companies Mislead Doctors and Harm Patients*. London: HarperCollins Publishers.
Hadler, Nortin. 2013. *The Citizen Patient: Reforming Health Care for the Sake of the Patient, Not the System*. Chapel Hill: The University of North Carolina Press.
———. 2004. *The Last Well Person: How to Stay Well Despite the Health-Care System*. Montreal: McGill-Queens University Press.
Heider, Fritz. 1958. *The Psychology of Interpersonal Relations*. New York: Wiley.
Hoffman, Martin L. 2000. *Empathy and Moral Development: Implications for Caring and Justice*. New York: Cambridge University Press.
Jones, W.H.S. 1923. *Hippocrates*. Cambridge: Harvard University Press.
Kelly, Kate, and Peggy Ramundo. 1993. *You Mean I'm Not Lazy, Stupid or Crazy? A Self-help Book for Adults with Attention Deficient Disorder*. Cincinnati: Tyrell and Severn.
Kohn, Linda T., J.M. Corrigan, and M.S. Donaldson, eds. 2000. *To Err Is Human: Building a Safer Health System*. Washington, DC: National Academy Press.
Landrigan, C.P. 2010. Temporal Trends in Rates of Patient Harm Resulting from Medical Care. *New England Journal of Medicine* 363 (22): 2124–2134.
Levenson, Robert W., and Ann M. Ruef. 1992. Empathy: A Physiological Substrate. *Journal of Personality and Social Psychology* 63: 234–246.
Lifton, Robert Jay. 1986. *The Nazi Doctors: Medical Killing and the Psychology of Genocide*. New York: Basic Books.
Makary, Marty. 2012. *Unaccountable: What Hospitals Won't Tell You and How Transparency Can Revolutionize Health Care*. New York: Bloomsbury Press.
Meunnig, Peter A., and Sherry A. Glied. 2010, November. What Changes in Survival Tell Us About US Health Care. *Health Affairs* 29 (11).
Mello, Michelle H., Amitabh Chandra, Atul A. Gawanda, and David M. Studdert. 2010, September. National Costs of the Medical Liability System. *Health Affairs* 29 (9): 1569–1577.
Mill, John Stuart. 1961. Utilitarianism. In *The Philosophy of John Stuart Mill*, ed. Marshall Cohen. New York: Random House.
Niebuhr, H. Richard. 1963. *The Responsible Self*. New York: Harper and Row.
Nussbaum, Martha C. 2006. *Frontiers of Justice: Disability, Nationality, Species Membership*. Cambridge, MA: Harvard University Press.
———. 2000. *Women and Human Development: The Capabilities Approach*. New York: Cambridge university Press.
Pellegrino, Edmund. 1998. Rationing Health Care: The Ethics of Medical Gatekeeping. In *Health Care Ethics: Critical Issues for the 21st Century*, ed. John F. Monagle and David C. Thomasma. Gathersburg: Aspen Publishers.
Perlmutter, David, and Carol Colman. 2004. *The Better Brain Book: The Best Tools for Improving Memory and Sharpness and for Preventing Aging of the Brain*. New York: Riverhead Books.
Rawls, John. 2001a. *A Theory of Justice*. Cambridge, MA: Harvard University Press.
———. 2001b. *Justice as Fairness: A Restatement*. Cambridge, MA: Harvard University Press.
Reiser, Stanley Joel. 2009. *Technological Medicine: The Changing World of Doctors and Patients*. New York: Cambridge University Press.

Relman, Arnold S. 2007. *A Second Opinion: Rescuing America's Health Care: A Plan for Universal Coverage Serving Patients Over Profit, The Century Foundation.* New York: Public Affairs.
Reuters, Doctors, Nurses Among Hundreds Charged with Defrauding Health Programs, New Hampshire Union Leader, July 14, 2017, p. A8.
Ross, W.D. 1930. *The Right and the Good.* Cambridge, MA: Harvard University Press.
Sen, Amartya. 2009. *The Idea of Justice.* Cambridge, MA: Harvard University Press.
Schore, Alan N. 1994. *Affect Regulation and the Origin of the Self: The Neurobiology of Emotional Development.* Hillsdale: Lawrence Erlbaum Associates.
Siwek, Jay. 2012, July 15. Choosing Wisely: Top Interventions to Improve Health and Reduce Harm, While Lowering Costs. *American Family Physician* 86 (2): 128–133.
Simons, Menno. 1972. The Thirst for Profit. In *Readings from Mennonite Writings: New and Old*, ed. I. Craig Haas. Intercourse: Good Books.
Solomon, Dave, Market Basket Protest: A Year Later, New Hampshire Sunday News, July 5, 2015, pages A1 and A4.
Thomas, William J., Erika C. Ziller, and Deborah Thayer. Low Costs of Defensive Medicine, Small Savings from Tort Reform. *Health Affairs* 29 (9): 1578–1584.
Veatch, Robert M. 2012. *Hippocratic, Religious and Secular Medical Ethics: The Points of Conflict.* Washington, DC: Georgetown University press.
Welch, H. Gilbert, Lisa M. Schwartz, and Steven Woloshin. 2011. *Over-diagnosed: Making People Sick in the Pursuit of Health.* Boston: Beacon Free Books.

Index

A
Abramson, J. MD., 93–96, 98–102, 104–106, 110, 116–118, 127, 137, 142, 144, 145, 149, 152, 153, 165, 166, 169, 178, 185, 189, 191, 193, 195
Accountability for reasonableness, 48, 67, 78
Affordable Care Act (ACA), 181
American Medical Association (AMA)
  codes of ethics
    1903, 1
    1980, 2, 14, 183
Angell, M. MD., 148
Applegate, W. MD., 196
Aristotle, 37
Augustine, 19

B
Brill, S., ix, 158, 181, 183, 186

C
Capabilities, 54–64, 68–71, 75–77, 88
Chernew, M., 84, 85
Choosing Wisely, viii, 88–90, 93, 97, 106, 109, 112, 135, 168, 187, 190, 200
Compassion, vii, 8–10, 16, 26, 33, 34, 36, 39, 48, 50, 54, 58, 65, 84, 145, 185, 200
CT scan, ix, 110, 160, 186
Cultural relativism, 10

D
Daniels, N., 2, 41–43, 45–56, 59, 61, 62, 64–68, 73–81, 83, 84

Defensive medicine, ix, 165, 166, 186–188
Dignity, 54, 55, 58, 60, 61, 70, 71, 76, 77
Disease managing, 140
Drug companies
  clinical standards, 145
  continuing education, 144, 146
  corrupting and biasing, 144–150
  deceptive bundling, 140
  direct consumer ads, 120, 152, 154, 170, 171
  disease monitoring, 180
  drug representatives, 144
  government facilitation, 155–157
  high prices, 154
  ineffective testing, 137–138
  medical journals, vi
  MGH pain centers, 146
  reporting misleading results, 141
  rigged research, viii
  stopping research trials early, 138–139
  testing outcomes to get a positive result, 139–140
  too short research trials, 138
  unjustifiable practices, v

F
Firth, R., 29, 30, 33, 34, 38, 40
Food and Drug Administration (FDA)
  approving faulty research, 170
  approving harmful drugs, 162
  not withdrawing harmful drugs, 163, 185
  omitting side effects, 153
  protecting high prices, 172
  suppressing scientific freedom, 163, 184

Food and Drug Administration (FDA) (*cont.*)
  unjustifiable practices, 170, 176
  withholding knowledge, 162
Frances, A., 18, 113–116, 118–133, 136, 151, 152, 154, 170, 192–195
Freedoms, 61, 63, 71–73
Frost, R., v–x, 168, 185, 203

G
Glendon, M.A., 6, 189
Goldacre, B. MD., 137, 138, 142–144, 149, 154, 164, 169, 175, 178
Golden Rule, 32, 33, 36–39, 75, 77
Good Samaritan, 6, 36, 38, 183
Gouging, 186
Graham, D., 143, 176, 177
Grundy, S. MD., 105
Gueriguian, J.L. MD., 177

H
Hadler, N.M. MD., v, 97, 135, 164, 167, 188, 201
Hallowell, E. MD., 121
Harm, vi, 3, 23, 43, 84, 117–119, 136, 168
Health, v, 2, 25, 41, 84, 116, 135, 167
Heider, F., 27, 28
Hippocrates, vi, 1, 11, 12, 17–20, 39, 41, 60, 179, 199
Hippocratic, vi, 1, 2, 11–14, 17–21, 23, 24, 39–41, 43, 44, 52, 54, 71, 73, 76, 80, 83, 84, 89, 159, 161, 166, 170, 182, 183, 185, 193, 199, 202
Hoffman, L.M., 8, 10, 43

I
Ideal observer, 29, 34, 38
Impartiality, 29, 30, 33–40, 74, 77, 150, 180
Inhibitions, 5, 7–11, 14, 25, 26, 34, 35, 43, 65, 199, 200

J
Jesus, 36, 37
Jones, W.H.S., 13, 43
Justice
  demands of, vi–viii, x, 2, 3, 8, 10, 13, 16, 18–21, 41, 45, 47, 50–53, 56, 57, 61, 64–66, 73, 76, 112, 166, 180, 199, 200
  distributive, 43, 45, 46, 48, 51

fairness, 44–54, 65
generic, 3, 11, 23, 43–45, 47, 51, 52, 54, 71–73, 77, 83, 84, 199
injustice, v–x, 3, 11, 38, 39, 44, 62, 64, 72, 74, 87, 166, 168, 185
unjust, x, 10, 36, 50, 55, 66, 70, 75, 76, 135, 168, 199, 203

K
Kant, I., 29

L
Law
  legal enforcement, 4, 79
  protection, 57
Life expectancy, viii, 86
Lifton, R.J., 28, 52
Logic, 25, 31, 90

M
Malpractice
  insurance, ix, 165, 166, 187
Markary, M. MD., 182, 196, 197
Mead, G.H., 29
Mediterranean diet, 105, 106, 191
Mercy, v–x, 133, 135, 160, 168, 182, 183, 185, 203
Mill, J.S., 3, 4, 59
Minimum medical care, 41–81
Muenning, P., 85–87

N
Nature, vi, 8–10, 13, 14, 33, 37, 59, 62, 63, 132, 145
Neuroscience, vi, vii, 8, 10, 11, 36, 43
Niebuhr, H.R., 29, 38
Nussbaum, M.C., 2, 41–43, 54–64, 68–81, 83, 84

O
Obama, B., 39, 158, 161
Objective, 28, 29, 31, 33, 35, 38, 40, 62, 180, 198
Over-diagnosis of behavior and feelings
  attention deficit disorder (ADHD), 119
  autism, 125, 126
  bi-polar disorder, 124, 125
  cases of diagnostic inflation, viii

depression, 115–119
diagnostic and statistical manual of mental disorders (DSM), 114, 115
inflation of mental disorders (illnesses), 113
medicalization normality—new categories (DSM-5), 113
new categories
  behavioral addictions, 132, 194
  the forgetting of normal aging (MNCD), 130, 131, 194
  gluttony (BED), 129, 131, 194
  mourning (MDD), 129, 132, 194
  temper tantrums (DMDD), 129, 130, 194
  Social anxiety disorder (SAD), 127, 128
Over-diagnosis of physical conditions
  cardiac interventions, 110–111
  colorectal screening, 110
  diabetes-2, 106–108
  heart disease, 97–106
  hypertension, 109
  infant intensive care, 111
  monitoring of pregnant women, 111–112
  osteoporosis, 91–96
  prostate cancer, 109
  samples of cases, 116
  should be curbed, 89
  thyroid cancer, 110

P
Pellegrino, E. MD., 53
Perlmutter, D. MD., 103
Physical fitness, 104, 106
Policy, vii, ix, x, 2, 3, 10, 24, 25, 30–33, 40, 42, 46, 53–55, 57, 59–61, 63, 64, 66, 70–72, 74–80, 84, 90, 93, 94, 99, 106, 131, 133, 135–203
Profit(s), v, ix, x, 1, 112, 114, 122, 131, 136, 139, 147, 149, 153–156, 158–161, 170, 173, 176, 177, 181–185, 189, 201–203

R
Rawls, J., 36, 45, 46, 54, 60, 62, 65
Reiser, S. MD., 1, 14–16, 57
Relman, A. MD., 136
Research, vi, viii, xi, 8, 9, 18, 29, 39, 44, 58, 93–97, 102, 104, 106–110, 118, 129, 136–150, 153, 154, 156, 162, 164, 166, 169, 173–175, 177–179, 181, 183, 189, 195–197

Rights, v, 3, 23, 60, 83, 155
Rotey, J. MD., 121

S
Science, vi, vii, 8, 10, 13, 29–31, 36, 40, 43, 136, 143, 147, 157, 161, 165, 175, 176, 185, 195
Sen, A., 2, 35, 41, 42, 54–64
Sensory observations, 25
Sherry, A.G., 51, 85
Side effects, 13, 85, 91–98, 100, 102–104, 106, 108, 110, 114, 118, 122, 123, 125, 128–131, 133, 136, 138, 141–143, 152–154, 157, 162, 164, 170, 172, 176, 177, 191–193, 196
Simons, M., v, 203
Siwek, J. MD., 88–90, 93, 97
Smith, A., 29, 35, 38, 74
Solomon, A., 26, 183
Statins, 97–106
Suggested healthcare policy changes
  government policies to control high prices
    invite profit sharing, 155
    negotiate prices, 155
    institute excise taxes for competition from foreign firms, 122
    eliminate price gouging, 184
    patient care, 155
    reduce patenting, 185–186
    tort reform, out of court negotiations, 163
  physicians seeking high moral ground
    alternatives to unnecessary diagnoses and drug use, 193
    avoiding medicalizing the normal, 115
    develop advisory committees, 15, 178, 196
    ethical rules for conflict of interest, 150, 164, 165, 179
    facilitate a teamwork model, 197–199
    reduce medical errors, 197–199
    stepped diagnoses, 192–193
  reform the F.D.A.
    assure accurate, available science, 161, 168, 175
    come down hard on deceptive drug ads, 170–172
    disallow some drug approval processes, 169–170
    eliminate conflicts of interest, 175–180
    refrain from facilitating the high price of drugs, 172–174

Suicide, 37, 117, 118, 128, 132, 140, 152, 171, 192

T
Tempkin, O. MD., 11, 12, 19, 20, 39

U
Universal healthcare, 21
Unsustainability, x, 81, 84, 135, 181, 200

V
Veatch, R., 17–20, 43
Vioxx, 142, 143, 147, 149, 150, 153, 170, 172, 176

W
Welch, G.H. MD., 90–93, 95, 99–101, 107, 108, 111, 165, 186, 190, 191
Westermark, E., 29
Willett, J.B., 105